£6·50

M. MARTIN

REVISE
THROUGH
DIAGRAMS

CHEMISTRY

Michael Lewis

Oxford University Press

Oxford University Press,
Great Clarendon Street, Oxford OX2 6DP

Oxford New York
Athens Auckland Bangkok Bogota Bombay
Buenos Aires Calcutta Cape Town Dar es Salaam Delhi
Florence Hong Kong Istanbul Karachi
Kuala Lumpur Madras Madrid Melbourne
Mexico City Nairobi Paris Singapore
Taipei Tokyo Toronto
and associated companies in
Berlin Ibadan

Oxford is a trade mark of Oxford University Press

© **Michael Lewis**

First published 1997

ISBN 0 19 914188 6 (Student's edition)
        0 19 914189 4 (Bookshop edition)

Typesetting, design and illustration by Hardlines, Charlbury, Oxford
Printed in Great Britain by Ebenezer Baylis and Son Ltd,
The Trinity Press, Worcester, and London

CONTENTS

CLASSIFYING MATERIALS:
what are things made of?

PATTERNS OF BEHAVIOUR
IN CHEMISTRY:
all about chemical reactions

CHANGING MATERIALS:
using Earth's resources

CHEMICAL CALCULATIONS

PRACTICAL SECTION

INDEX

Matter and energy

MATTER

Matter makes up all the substances around us.

Matter can be recognized by weighing because it has mass.

Matter exists in three **states**: **solid**, **liquid**, and **gas**.

When the particles of a substance gain or lose energy, the substance may change state.

Melting and boiling points

These are **temperatures at which pure substances change state**.

They are used by chemists to:
- recognize substances: no two substances have exactly the same melting or boiling points;
- check purity: impurities change melting and boiling points.

Sublimation

This is the process of turning directly from solid to gas (instead of melting to a liquid).

Iodine and carbon dioxide (dry ice) both sublime.

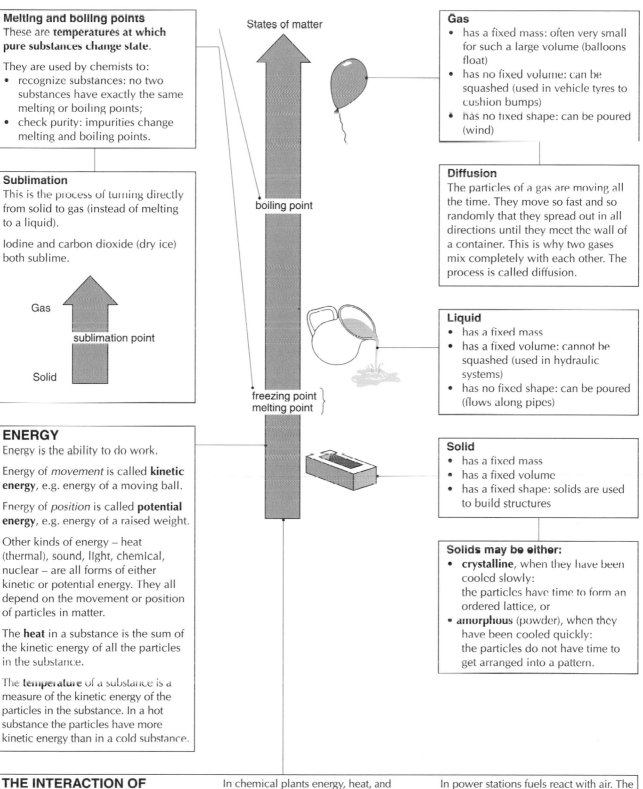

Gas

sublimation point

Solid

ENERGY

Energy is the ability to do work.

Energy of *movement* is called **kinetic energy**, e.g. energy of a moving ball.

Energy of *position* is called **potential energy**, e.g. energy of a raised weight.

Other kinds of energy – heat (thermal), sound, light, chemical, nuclear – are all forms of either kinetic or potential energy. They all depend on the movement or position of particles in matter.

The **heat** in a substance is the sum of the kinetic energy of all the particles in the substance.

The **temperature** of a substance is a measure of the kinetic energy of the particles in the substance. In a hot substance the particles have more kinetic energy than in a cold substance.

States of matter

boiling point

freezing point
melting point

Gas
- has a fixed mass: often very small for such a large volume (balloons float)
- has no fixed volume: can be squashed (used in vehicle tyres to cushion bumps)
- has no fixed shape: can be poured (wind)

Diffusion

The particles of a gas are moving all the time. They move so fast and so randomly that they spread out in all directions until they meet the wall of a container. This is why two gases mix completely with each other. The process is called diffusion.

Liquid
- has a fixed mass
- has a fixed volume: cannot be squashed (used in hydraulic systems)
- has no fixed shape: can be poured (flows along pipes)

Solid
- has a fixed mass
- has a fixed volume
- has a fixed shape: solids are used to build structures

Solids may be either:
- **crystalline**, when they have been cooled slowly: the particles have time to form an ordered lattice, or
- **amorphous** (powder), when they have been cooled quickly: the particles do not have time to get arranged into a pattern.

THE INTERACTION OF MATTER AND ENERGY

In chemistry we study the effects of applying energy to matter or getting matter to release stored energy.

In chemical plants energy, heat, and pressure are applied to raw materials like crude oil or iron ore, changing them into useful substances that we need.

In power stations fuels react with air. The potential energy stored in the fuel (in the bonds between the atoms) is released as kinetic (heat) energy which is made to do work generating electricity.

Changing state and the kinetic theory

① When a gas is cooled, its temperature falls until it reaches its boiling point.
② It then begins to condense into a liquid.
③ The temperature does not drop any further until all the gas has condensed.
④ The liquid then cools until it reaches its freezing point.
⑤ It then begins to freeze, forming a solid.
⑥ The temperature does not drop any further until all the liquid has frozen.

This behaviour is shown here on a **cooling curve**, which shows how temperature changes with time. The behaviour is explained using the **kinetic theory**.

The kinetic theory explains the properties of the different states of matter by the behaviour of the particles in the matter. It describes the movement of the particles (the **kinetic energy**) and their relative positions (the **potential energy**).

COOLING CURVE FOR WATER

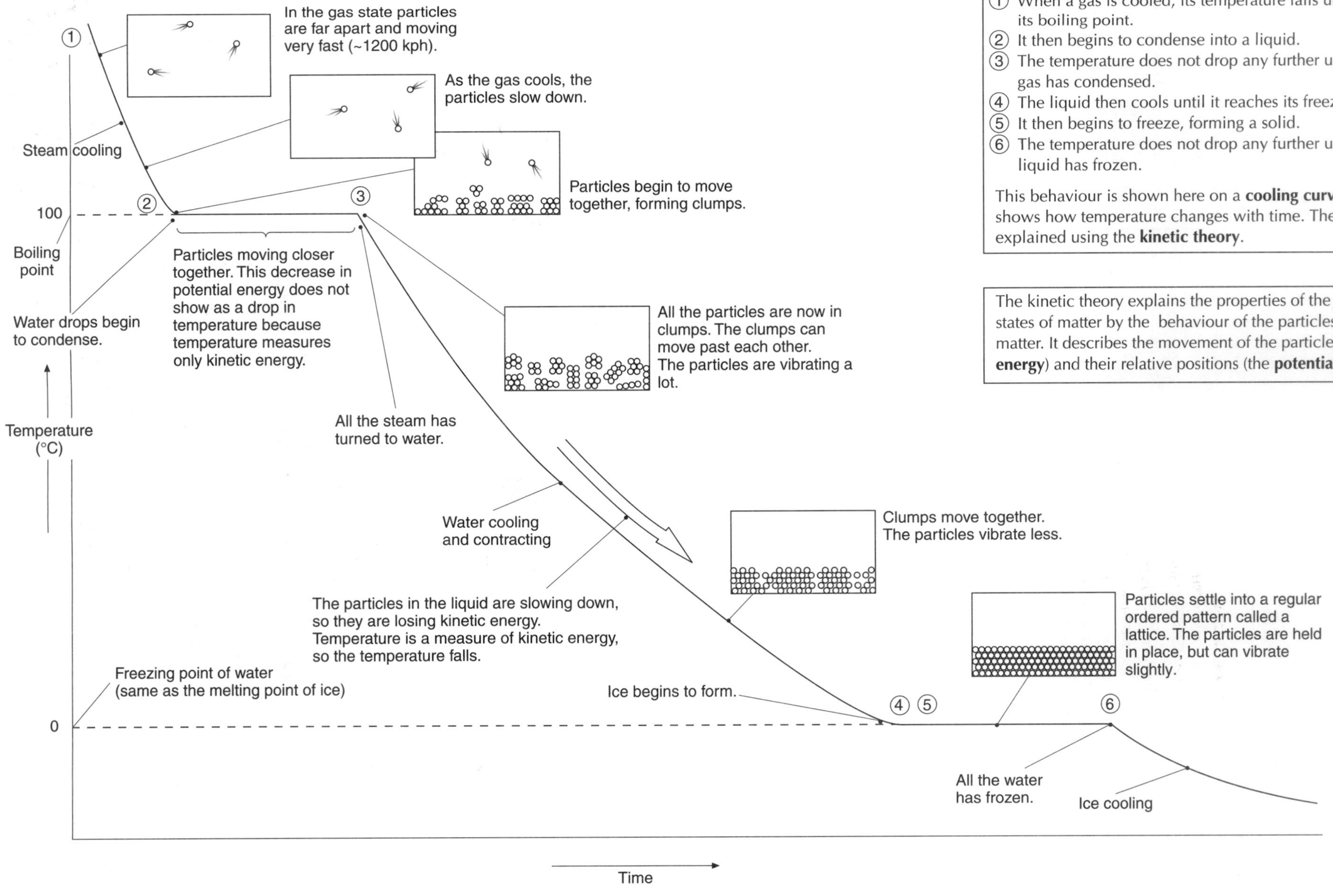

In the gas state particles are far apart and moving very fast (~1200 kph).

As the gas cools, the particles slow down.

Particles begin to move together, forming clumps.

All the particles are now in clumps. The clumps can move past each other. The particles are vibrating a lot.

Clumps move together. The particles vibrate less.

Particles settle into a regular ordered pattern called a lattice. The particles are held in place, but can vibrate slightly.

Particles moving closer together. This decrease in potential energy does not show as a drop in temperature because temperature measures only kinetic energy.

All the steam has turned to water.

Water cooling and contracting

The particles in the liquid are slowing down, so they are losing kinetic energy. Temperature is a measure of kinetic energy, so the temperature falls.

Ice begins to form.

Freezing point of water (same as the melting point of ice)

All the water has frozen.

Ice cooling

Steam cooling

Water drops begin to condense.

100

Boiling point

Temperature (°C)

0

Time

6 Changing state and the kinetic theory

Elements, mixtures, and compounds

ELEMENTS

- cannot be decomposed
- made of only one kind of atom
- two main kinds: metals and non-metals

Metals

Physical properties
- conduct electricity
- ductile and malleable (can be bent and shaped)

Chemical properties
- form basic oxides, e.g. MgO; CuO
- form cations, e.g. Na^+; Ca^{2+}

Non-metals

Physical properties
- insulators
- brittle (suddenly snap when loaded)

Chemical properties
- form acidic oxides, e.g. CO_2; SO_2; NO_2
- form anions, e.g. O^{2-}; Cl^-; Br^-

Relative amounts in Earth's crust

Element	%	
oxygen	46.6	most abundant non-metal
silicon	27.7	
aluminium	8.1	most abundant metal
iron	5.0	
calcium	3.6	
hydrogen	0.22	
carbon	0.19	forms most compounds

IRON FILINGS
grey, magnetic

SULPHUR
yellow, low m.p.

MIXTURES

- made by physical change
- original properties still remain
- have variable composition
- there is no energy change when mixing
- separated by physical changes
- made from elements or compounds

MIXTURE
Both grey iron filings and yellow sulphur can still be seen; separate with magnet.

Important examples

Air
- mixture of gases
- main components

nitrogen	78%
oxygen	21%
noble gases	0.9%
carbon dioxide	0.04%

- separated by fractional distillation

Petroleum
- mixture of saturated hydrocarbons
- composition changes from one source to another
- separated by fractional distillation

Sea water
- mixture of ionic salts in water

Salt	Mass (g in 100g of sea water)
sodium chloride	2.6
magnesium chloride	0.3
magnesium sulphate	0.2
calcium sulphate	0.1
potassium chloride	0.1

- separated by distillation to collect water, and evaporation to collect salt.

MIX

COMBINE

COMPOUNDS

- made by chemical change
- have new properties, different from reactants
- have fixed composition and definite formula
- there is an energy change during combination
- can only be separated by decomposition (a chemical change)

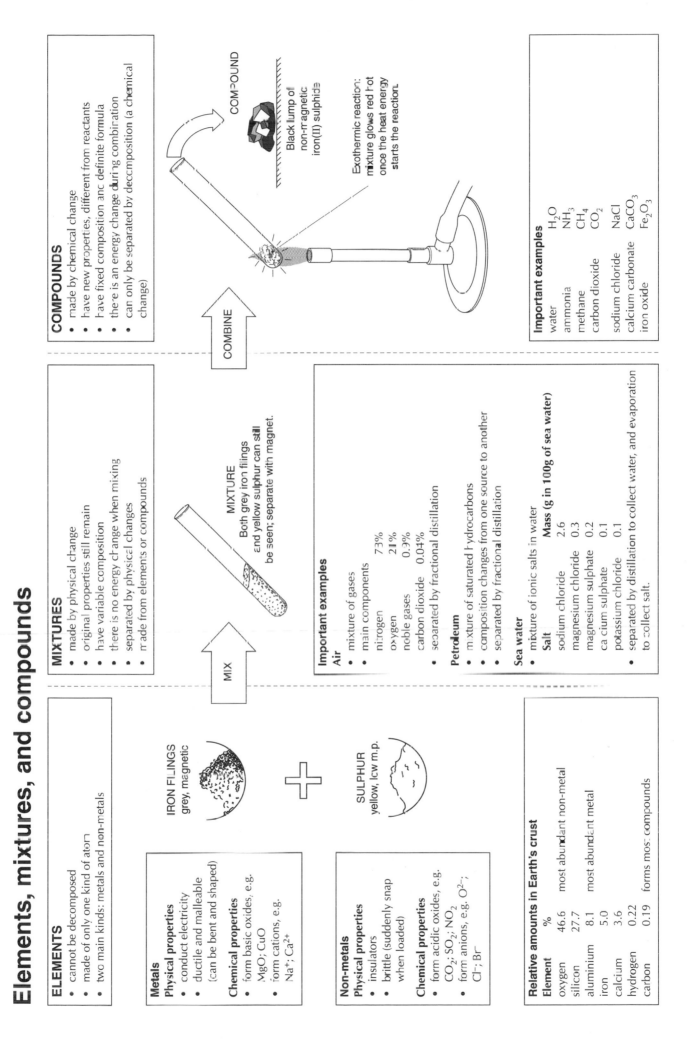

COMPOUND

Black lump of non-magnetic iron(II) sulphide

Exothermic reaction: mixture glows red hot once the heat energy starts the reaction.

Important examples

water	H_2O
ammonia	NH_3
methane	CH_4
carbon dioxide	CO_2
sodium chloride	NaCl
calcium carbonate	$CaCO_3$
iron oxide	Fe_2O_3

Mixtures

- made of more than one substance
- have variable composition
- properties are those of the original substances added together
- no energy change when mixing happens

SOLUTIONS

Recognized by being transparent – you can see through them.

The **solute** is the smaller part of the solution, often a solid.

The **solvent** is the larger part of a solution, usually a liquid.

Solutions are described as **homogeneous** or **uniform mixtures**.

Both the solute and solvent particles are very small and nearly the same size, so they are all in the same state or phase.

Important examples of solutions

Air

A mixture of gases. Nitrogen (~78%) is the solvent. Oxygen (~21%), the noble gases, and carbon dioxide are some of the solutes. The b.p.s are very close, so air is separated by fractional distillation:

nitrogen boils at $-196\,°C$
oxygen boils at $-183\,°C$

Nitrogen boils first and comes off the top of the column. Oxygen is left at the bottom.

Crude oil

A mixture of saturated hydrocarbons. Composition varies and all the boiling points are very close together. So it is separated into **fractions**, which are groups of hydrocarbons with similar boiling points.

Sea water

Sea water is evaporated to provide common salt (sodium chloride) and the other solutes it contains. In dry parts of the world it is used to provide water for drinking and irrigation. The sea water is distilled, or separated by a process called reverse osmosis.

SUSPENSIONS

Recognized by being opaque – you cannot see through them.

Here the particles which are suspended are much larger than the particles they are suspended in. The suspended particles are actually clumps of pure solid or liquid. They are big enough to reflect light, so the mixture is opaque. Because these big particles are in a different state from the smaller particles, suspensions are called **heterogeneous** or **non-uniform mixtures**.

Often the larger particles sink to the bottom and the suspension separates by itself, but this can take a long time.

Important examples of suspensions

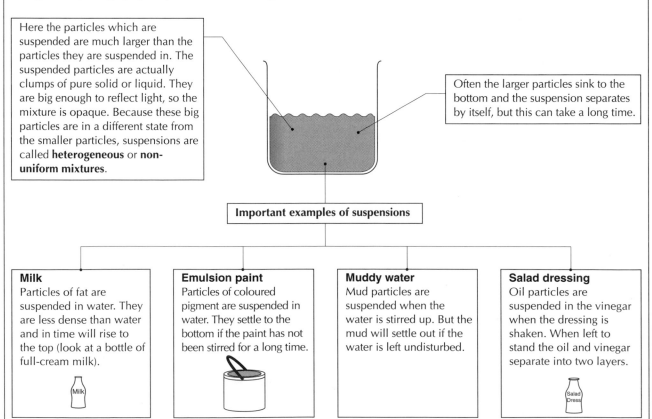

Milk
Particles of fat are suspended in water. They are less dense than water and in time will rise to the top (look at a bottle of full-cream milk).

Emulsion paint
Particles of coloured pigment are suspended in water. They settle to the bottom if the paint has not been stirred for a long time.

Muddy water
Mud particles are suspended when the water is stirred up. But the mud will settle out if the water is left undisturbed.

Salad dressing
Oil particles are suspended in the vinegar when the dressing is shaken. When left to stand the oil and vinegar separate into two layers.

Separating mixtures

SOLUTIONS

Because all the particles are in the same state, solutions can only be separated by making one part change state.

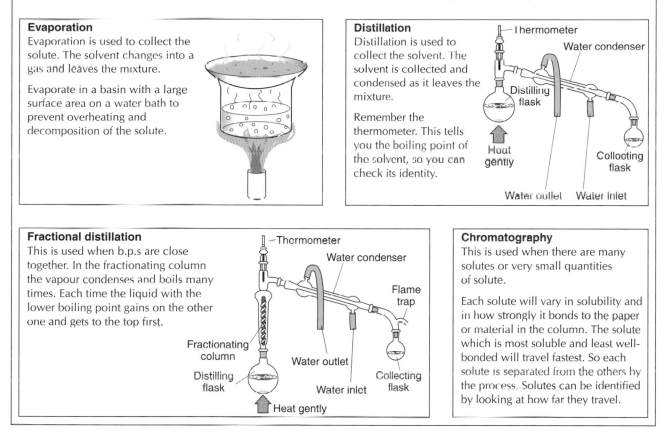

Evaporation

Evaporation is used to collect the solute. The solvent changes into a gas and leaves the mixture.

Evaporate in a basin with a large surface area on a water bath to prevent overheating and decomposition of the solute.

Distillation

Distillation is used to collect the solvent. The solvent is collected and condensed as it leaves the mixture.

Remember the thermometer. This tells you the boiling point of the solvent, so you can check its identity.

Thermometer
Water condenser
Distilling flask
Heat gently
Collecting flask
Water outlet   Water inlet

Fractional distillation

This is used when b.p.s are close together. In the fractionating column the vapour condenses and boils many times. Each time the liquid with the lower boiling point gains on the other one and gets to the top first.

Thermometer
Water condenser
Flame trap
Fractionating column
Water outlet
Distilling flask
Water inlet
Collecting flask
Heat gently

Chromatography

This is used when there are many solutes or very small quantities of solute.

Each solute will vary in solubility and in how strongly it bonds to the paper or material in the column. The solute which is most soluble and least well-bonded will travel fastest. So each solute is separated from the others by the process. Solutes can be identified by looking at how far they travel.

SUSPENSIONS

Suspensions are usually easier to separate than solutions because the suspended particles are in a different state. Suspensions can be separated by trapping the larger particles. This is done:

- in filter paper or in a bed of sand
- by careful decanting
- using a tap or separation funnel

Centrifuging

If a suspension is spun very fast, the larger particles move to the outside. So suspensions can be separated by suspending tubes in a centrifuge and spinning them. The residue packs down at the bottom of the tube and the filtrate (liquid) can be poured off.

Filtering

Suspended particles are trapped on the filter paper forming the **residue**. The smaller particles pass through the filter paper forming the **filtrate**.

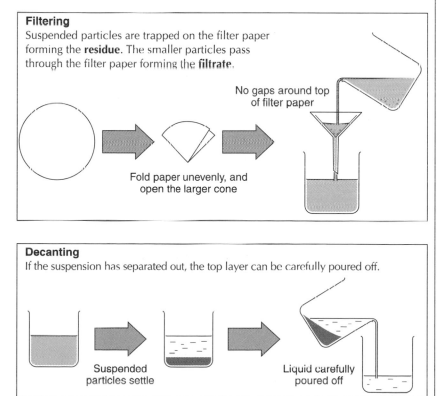

No gaps around top of filter paper

Fold paper unevenly, and open the larger cone

Using a tap funnel

Two insoluble (**immiscible**) liquids can be separated by running off the lower layer.

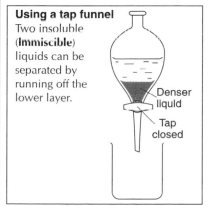

Denser liquid
Tap closed

Decanting

If the suspension has separated out, the top layer can be carefully poured off.

Suspended particles settle

Liquid carefully poured off

Atomic structure

The nuclear atom is made of a central part called the nucleus surrounded by an outer part.

NUCLEUS

- very *small*: less than 0.1% of the volume
- very *massive*: contains 99.9% of the matter
- so *very, very dense*

Contains particles called:

    protons: relative mass 1 unit; relative charge +1 unit

and **neutrons**: relative mass 1 unit; charge 0

Atomic number

The atomic number of an element is the *number of protons in the atoms of that element.*

Mass number

The mass number of an atom is the *sum of the protons and neutrons in the nucleus of an atom.*

All the atoms of an element have the same atomic number.

ISOTOPES

The number of neutrons in the atoms of an element can vary.

Atoms with the same atomic number but different mass numbers are called **isotopes**.

e.g. there are three isotopes of hydrogen

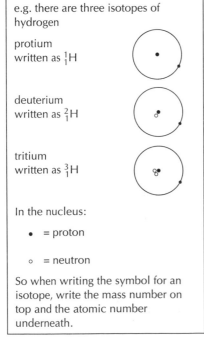

protium written as $^1_1$H

deuterium written as $^2_1$H

tritium written as $^3_1$H

In the nucleus:

- • = proton

- ○ = neutron

So when writing the symbol for an isotope, write the mass number on top and the atomic number underneath.

THE OUTSIDE

- makes up 99.9% of the volume of the atom
- contains *hardly any mass* (about 0.1%), so is mainly empty space
- is *negatively charged*

Contains particles called **electrons**.

Electrons have very little mass – about 1840 times less than a proton or neutron.

Electrons are negatively charged with a relative charge of –1 unit.

The charge on one electron cancels out the charge on one proton.

Shells

Electrons are arranged around the nucleus in groups called shells.

The shells are numbered starting at the centre and working outwards.

The maximum number of electrons in each shell differs:

    the *first* shell can hold up to **2**
    the *second* shell can hold up to **8**
    the *third* shell can hold up to **8**

ELECTRONIC STRUCTURE

The energy of the electrons in each shell increases with the distance from the nucleus. So electrons in the inner shell have less energy than those in the shells outside. The electrons in each atom go into the shells with the lowest energy, so the inner shells are always filled up first.

The electronic structure (**electronic configuration**) of an atom is the arrangement of the electrons around the nucleus of the atom. It is shown using either a diagram or a list of numbers.

It is worked out like this:

1. Look up the atomic number: this gives you the number of protons. In the neutral atom, the number of electrons is the same as the number of protons.

2. Start filling the shells until the right number of electrons have been placed.

e.g. **fluorine, atomic number 9**
so 2 electrons in the first shell
and 7 electrons in the second shell
written as **2.7**
drawn as

e.g. **magnesium, atomic number 12**
so 2 electrons in the first shell
and 8 electrons in the second shell
and 2 electrons in the third shell
written as **2.8.2**

drawn as

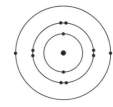

Atomic structure and the periodic table

The arrangement of electrons in an atom – the electronic structure or electronic configuration – is related to the position of the element in the periodic table. The electronic structure for the first three rows of elements is shown below.

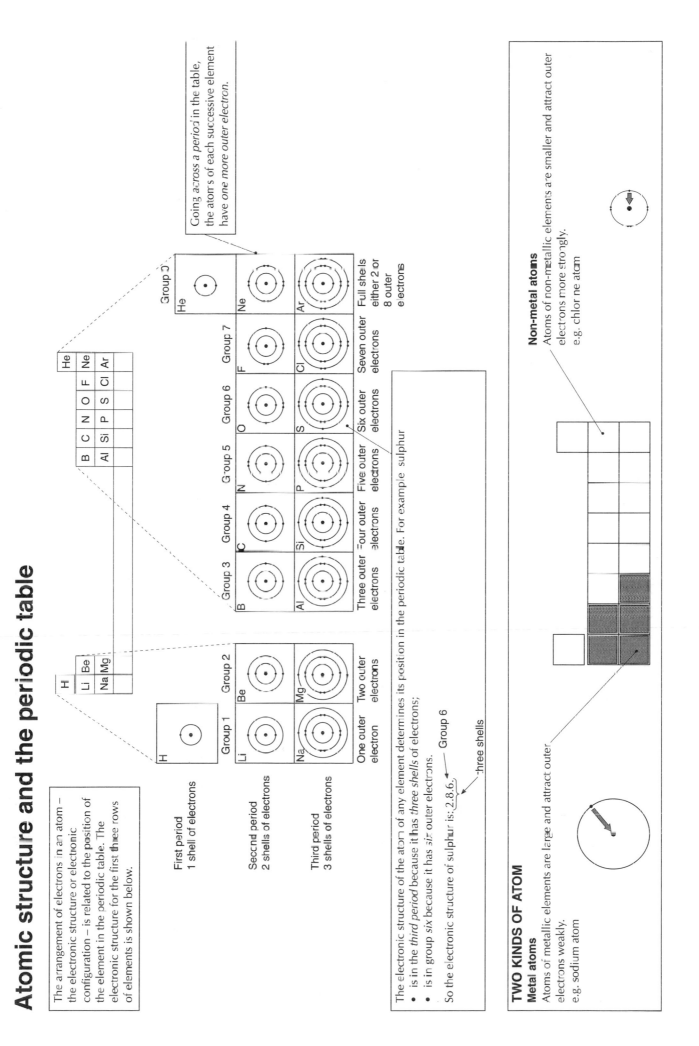

Going across a period in the table, the atoms of each successive element have *one more outer electron*.

First period
1 shell of electrons

Second period
2 shells of electrons

Third period
3 shells of electrons

Group 1	Group 2		Group 3	Group 4	Group 5	Group 6	Group 7	Group 0
H								He
Li	Be		B	C	N	O	F	Ne
Na	Mg		Al	Si	P	S	Cl	Ar
One outer electron	Two outer electrons		Three outer electrons	Four outer electrons	Five outer electrons	Six outer electrons	Seven outer electrons	Full shells either 2 or 8 outer electrons

The electronic structure of the atom of any element determines its position in the periodic table. For example sulphur

* is in the *third period* because it has *three shells* of electrons;
* is in group *six* because it has *six* outer electrons.

So the electronic structure of sulphur is: 2.8.6.

Group 6

three shells

TWO KINDS OF ATOM

Metal atoms
Atoms of metallic elements are large and attract outer electrons weakly.
e.g. sodium atom

Non-metal atoms
Atoms of non-metallic elements are smaller and attract outer electrons more strongly.
e.g. chlorine atom

Introduction to bonding and structure

BONDING

A **bond** is any attractive force between particles. In every bond, positive particles are attracting negative ones.

If electrons are lost by one atom and gained by another, then the two atoms become oppositely charged and a bond forms.

If electrons are shared between two atoms, then the protons in both atoms attract the shared electrons and another kind of bond forms.

The periodic table helps us decide on the kind of bond

METALS

NON-METALS

If both atoms are *metallic*, the bond is a *metallic* bond, e.g. in copper or aluminium.

If one atom is *metallic* and the other is *non-metallic*, the bond is *ionic*, e.g. in sodium chloride.

If both atoms are *non-metallic*, the bond is *covalent*, e.g. in carbon dioxide, water, or iodine.

Some compounds like sodium hydroxide have two types of bonding: ionic bonding between the sodium ion and the hydroxide ion, and covalent bonding between the hydrogen and oxygen atoms.

Bond strength

There is no simple rule about which types of bonds are strongest.

Metallic bonds vary from the weak ones in mercury to the very strong ones in tungsten.

Ionic and covalent bonds also vary in strength.

STRUCTURE

After bonding takes place, there are different kinds of particle:

Atoms

- Metals are made of giant lattices of atoms held together by metallic bonds.
- Carbon (diamond or graphite) lattices are made of giant lattices of atoms covalently bonded together.
- Noble gas lattices (only formed at very low temperatures) are made of atoms held together by weak van der Waals' forces.

Ions

Compounds made from metals and non-metals exist in giant lattices made of ions. Ions are atoms which have gained or lost electrons and so have a charge.

Cations (metal ions) have positive charges because they have lost electrons.

Anions (non-metal ions) have negative charges because they have gained extra electrons.

Molecules

Elements and compounds of non-metals exist as groups of atoms covalently bonded together. These are called **molecules**. Molecules, like atoms, have no charge.

A **diatomic molecule** is made from only two atoms, e.g. H_2; N_2; O_2; Cl_2; I_2.

A **triatomic molecule** is made from three atoms, e.g. H_2O; CO_2.

Polyatomic molecules are made from many atoms, e.g. CH_4; CH_3CH_2OH.

Bonding between molecules

Just as there are attractive forces between oppositely charged parts of next door atoms, there are also attractive forces between oppositely charged parts of next door molecules.

There are different kinds of forces between molecules. Some are called **polar forces** and some are called **van der Waals' forces**.

The most important thing to remember about these are:

- the forces inside molecules are much stronger than the forces between molecules
- the bigger the molecules are, the stronger the forces between them.

strong bond inside

weak force between

Bonding and structure in metals

A theory of bonding must explain the properties of the substances.

KEY METALLIC PROPERTIES

- good electrical conductivity
- malleable and ductile (can be bent and reshaped without suddenly snapping)

BONDING IN METALS

Remember that metal atoms are large and attract outer electrons weakly.

In a metal, next door atoms overlap because the outer part of the atoms is mostly empty space.

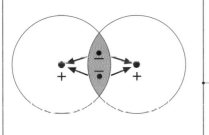

Both nuclei attract the electrons in the overlap area, so there is an attractive force between the atoms called a metallic bond.

Metallic bond: the electrostatic force of attraction that two neighbouring nuclei have for the shared electrons between them.

STRUCTURE IN METALS

In most metals the lattice of particles is arranged in the close-packed pattern shown below.

The lattice is formed of cations made of the nucleus and inner shell electrons. So the charge on the cations is always equal to the number of the group of the periodic table that the element is found in.

Between the cations there is a sea of outer electrons. These outer electrons are shared among all the atoms and can move anywhere in the lattice (they are **delocalized**).

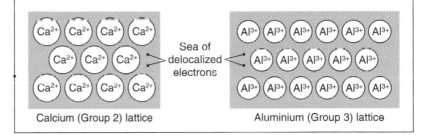

Calcium (Group 2) lattice   Aluminium (Group 3) lattice

EXPLAINING THE KEY PROPERTIES

Electrical conductivity

Conduction is the movement of charge. In metals the delocalized electrons drift through the metal lattice carrying the charge.

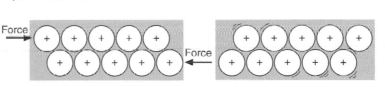

Malleability and ductility

The shape of metals can be changed because the layers of atoms can slide past or over each other. When they do this, some bonds are broken, but an equal number are made.

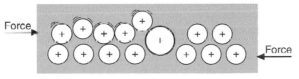

CHANGING METALLIC PROPERTIES

Alloying

An alloy is a mixture of a metal and another element. Adding another element disturbs the pattern in the lattice so that the layers will not slide past each other so easily. Alloys are usually stronger and harder than the pure metal.

Heat treatment

If a liquid metal is cooled quickly, the crystals or grains in it have little time to grow. But if it is cooled slowly, large crystals or grains can grow. The crystal or grain size affects the strength of the metal. Metals with larger grains are softer than metals with smaller grains.

If a solid metal is slowly heated to red heat (so that the atoms have enough energy to rearrange) and then is suddenly cooled, the atoms are trapped in a new pattern. The strength and hardness of the metal will be changed.

Uses of metals

The physical properties of metals make them suitable for particular purposes:

- **electrical conduction**: copper and aluminium cables
- **thermal or heat conduction**: copper bottoms to saucepans
- **malleability**: copper pipes, steel car bodies, lead roofs, etc.

Bonding and structure in ionic compounds

A theory of bonding must explain the properties of the substances.

KEY IONIC PROPERTIES

- hard, high melting point solids
- insulators when solid, but conduct when molten or dissolved
- brittle – they break suddenly under load

BONDING IN IONIC COMPOUNDS

Remember that metal atoms are large and hold outer electrons weakly, but non-metal atoms are small and hold outer electrons strongly.

This means that when metal and non-metal atoms meet, electrons are lost by the metal atom and gained by the non-metal atom.

The metal atom loses the outer electrons leaving a positive ion.

The number of electrons lost is the number of electrons in the outer shell. This is the same as the group number, so the charge on the positive ion equals the group number.

e.g. Group   1   2   3
          Na^+ Mg^{2+} Al^{3+}

The non-metal atom gains new outer electrons making a negative ion.

Electrons are gained from the metal atom until the outer shell is full. So the charge on the ion is the same as the number of gaps in the outer shell. This is given by (8 – group number).

e.g. Group   6   7   0
         O^{2-} F^- –

no gaps so
no ions formed

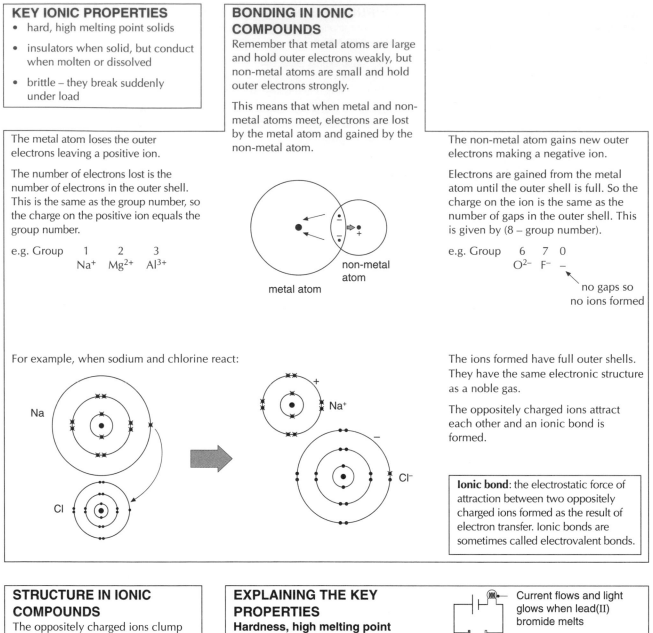

metal atom   non-metal atom

For example, when sodium and chlorine react:

Na   Cl Na^+ Cl^-

The ions formed have full outer shells. They have the same electronic structure as a noble gas.

The oppositely charged ions attract each other and an ionic bond is formed.

Ionic bond: the electrostatic force of attraction between two oppositely charged ions formed as the result of electron transfer. Ionic bonds are sometimes called electrovalent bonds.

STRUCTURE IN IONIC COMPOUNDS

The oppositely charged ions clump together and a lattice forms. When you draw an ionic lattice, do not draw lines between the ions: lines are used to represent covalent bonds. Remember that positive ions are nearly always smaller than negative ones.

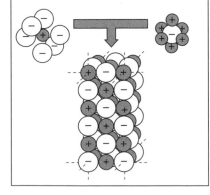

EXPLAINING THE KEY PROPERTIES

Hardness, high melting point
Oppositely charged ions attract each other strongly forming a lattice which is difficult to break up.

Electrical properties
In the solid state the ions are held in place and cannot move. Once the substance has melted or dissolved, the ions can move and carry charge. Positive ions are attracted towards the cathode: they are called cations. Negative ions are attracted towards the anode: they are called anions.

Brittleness
When a force is applied, the strongly bonded lattice resists. But eventually the layers start to slide past each other. Ions of the same charge move next to each other and repulsion replaces attraction. The solid breaks.

Current flows and light glows when lead(II) bromide melts

Lead(II) bromide   Anode   Cathode

Br^- Br^- Pb^{2+} Br^- Br^- Pb^{2+} e^-

Heat

Force   Force

Bonding in covalent substances

A theory of bonding must explain the properties of substances.

KEY COVALENT PROPERTIES

There are two different kinds of covalent substance:
- *either* gases, liquids, and soft solids, *or*
- hard, very high m.p. solids

Both are insulators.
Both are brittle in the solid state.

Covalent bonding is found in non-metallic elements as well as compounds.

BONDING IN COVALENT SUBSTANCES

Remember that non-metal atoms are smaller than metal ones and attract their electrons strongly.

When two non-metal atoms touch and overlap, pairs of electrons in the area of overlap get attracted to both nuclei and a bond forms.

Covalent bond: the electrostatic force of attraction that two nuclei have for a shared pair of electrons between them.

A covalent bond can be drawn in a number of ways. For example, the bond between two hydrogen atoms can be drawn in three ways.

H:H   or   H×H   or   H—H

NUMBER OF BONDS FORMED

The number of covalent bonds an atom of an element makes depends on the position of the element in the periodic table.

Remember that to make a covalent bond an atom needs
- an electron to put into the bond, and
- a space or gap in its outer shell into which the other atom's electron can go.

The table shows this.

Group	3	4	5	6	7	8
Number of electrons in outer shell	3	4	5	6	7	8
Number of gaps in outer shell	5	4	3	2	1	0
Number of bonds made	3	4	3	2	1	0
Examples	Cl. ×B:Cl Cl.	H H:C:H H H | H—C—H | H	H:N:H H (..) H—N—H | H	H:O: H H—O: | H	H:Cl: H—Cl:	:Ne:

All these molecules have one or more shared pairs of electrons making the bonds. Some of them also have pairs of electrons belonging to one atom only. These are called **non-bonding pairs** or **lone pairs**.

DOUBLE AND TRIPLE BONDING

Sometimes atoms can share more than one pair of electrons between them. When this happens, double or even triple bonds are made.

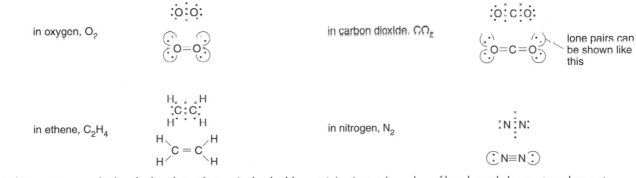

in oxygen, O_2

in ethene, C_2H_4

in carbon dioxide, CO_2

in nitrogen, N_2

lone pairs can be shown like this

It does not matter whether the bonds made are single, double, or triple: the total number of bonds made by an atom does not change. In the examples here, carbon still makes four bonds, nitrogen makes three, oxygen makes two, and hydrogen makes one.

Structure of covalent substances

MOLECULAR COVALENT SUBSTANCES

These substances exist as separate, covalently bonded molecules. Molecules have no charge.

The covalent bonding *inside* each molecule is *very strong*, but the bonding *between* the molecules is much, much *weaker*. It is this weak bonding which is overcome when these substances are broken or melted.

The solid lattices are made of ordered arrangements of molecules.

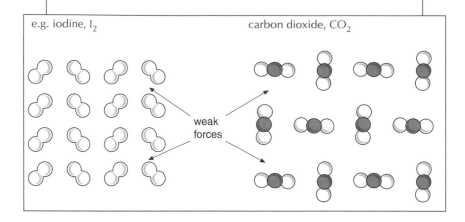

e.g. iodine, I_2

carbon dioxide, CO_2

weak forces

EXPLAINING THE KEY PROPERTIES

Softness and low melting point

When a force is applied to these substances, or when they are heated, it is the weak forces *between* the molecules that are overcome. The strong forces inside the molecules are not overcome.

Because these intermolecular (between molecules) forces are so weak, the solids are soft and melt easily.

Insulators

There are no charged particles free to move in these substances. The electrons are localized in pairs and there are no ions. So they do not conduct.

Brittleness

When force is applied and the weak forces are overcome, no new bonds are made. The substance suddenly breaks.

MACROMOLECULAR (GIANT COVALENT) SUBSTANCES

These substances exist in huge, covalently bonded lattices. A diamond is a single giant molecule with millions of bonds. Strong covalent bonding continues throughout the whole lattice. Here there are no weak forces to break, only strong covalent bonds.

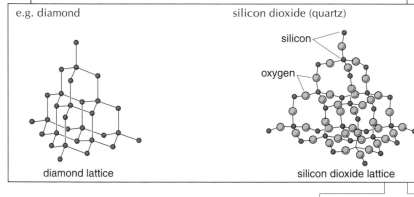

e.g. diamond

silicon dioxide (quartz)

silicon

oxygen

diamond lattice

silicon dioxide lattice

EXPLAINING THE KEY PROPERTIES

Hardness and high melting point

Breaking or melting these substances involves breaking strong covalent bonds. So these substances are very hard with very high m.p.s. Diamond is the hardest substance known.

Insulators

There are no charged particles free to move in these substances. The electrons are localized in pairs and there are no ions. So they do not conduct.

Brittleness

When force is applied and the weak forces are overcome, no new bonds are made. The substance suddenly breaks.

THE UNUSUAL PROPERTIES OF GRAPHITE

Graphite is a form of carbon. Although it is covalently bonded, it conducts electricity. Carbon rods used for electrolysis are made of graphite.

Graphite has a layered structure in which each carbon uses only three electrons for bonding. The fourth electron from each atom is delocalized and shared throughout the layer. Because graphite has delocalized electrons (like a metal), it conducts electricity (like a metal).

Carbon atoms also exist in cage-like lattices called fullerenes.

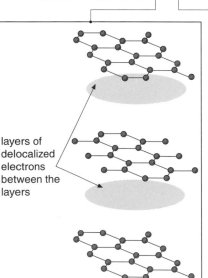

layers of delocalized electrons between the layers

POLYMERS

Polymers are giant chains of covalently bonded atoms forming extremely long molecules. Like other covalent molecules, the forces between them are much weaker than the forces inside them. But these long molecules can tangle together like string or spaghetti. This tangling is made worse if the polymer molecules have side chains sticking out of them.

Plastics are polymers. Some plastics are bendy and flexible because of this tangling of long polymer chains.

Tangled chains

Bonding and structure in the main groups of the periodic table

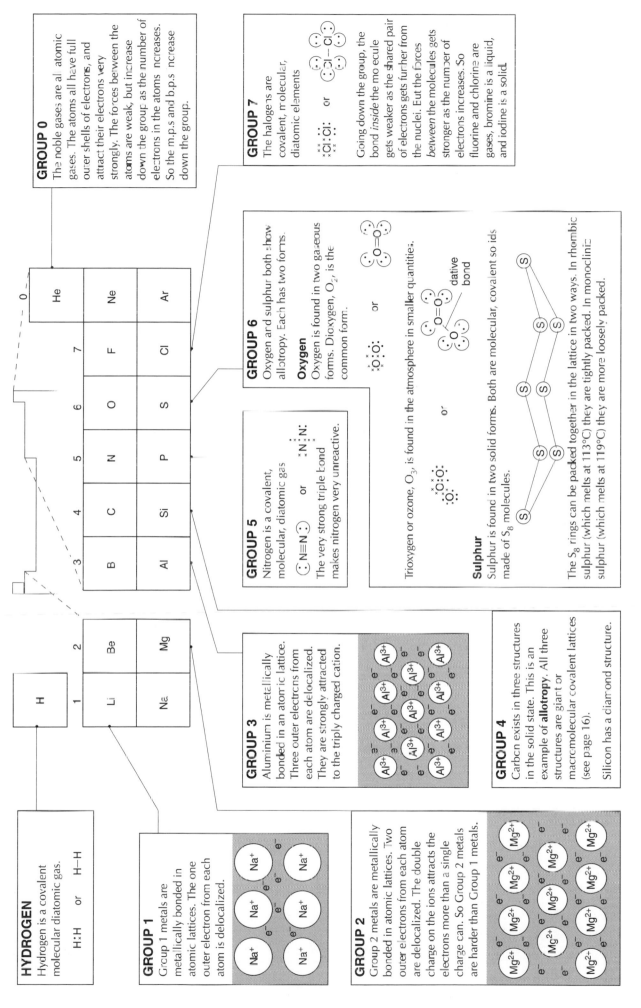

The periodic table

The periodic table is a list of the elements in order of their atomic numbers.

The list is set out so that elements with similar properties are in vertical columns.

GROUPS

- The long vertical columns of elements are called **groups**.
- The groups are numbered from 0 to 7.
- A group contains elements with similar properties.
- Going down a group, the elements show trends as the properties they share change slightly.

Hydrogen is the first element and is unlike any other. So hydrogen is in a box by itself and is not part of any of the groups.

GROUP NAMES

Some groups have names as well as numbers.

- Group 0 is called the **noble gases**.
- Group 1 is called the **alkali metals**.
- Group 2 is called the **alkaline earth metals**.
- Group 7 is called the **halogens**.

METALS AND NON-METALS

The simplest way to divide up the periodic table is into metallic and non-metallic elements.

The *metals* appear on the *left* of the zig-zag line, while the *non-metals* are on the *right* of the line. You can see that there are many more metals than non-metals.

METALLOIDS OR SEMI-METALS

Elements near the zig-zag line often show some metallic properties and some non-metallic properties. For example:

- carbon (graphite) is a non-metal, but conducts electricity
- silicon and germanium are called **semiconductors** and have resistances which change markedly with conditions
- aluminium oxide has both basic and acidic properties (**amphoteric**).

PERIODS

Periods are horizontal rows of elements. A period contains elements with different properties. Each period, apart from the first one, starts with metals and changes to non-metals.

— transition elements —

TRANSITION ELEMENTS

This is a large collection of elements between Groups 2 and 3. The elements in the main groups differ in properties from group to group. But the transition elements all have certain properties in common:

1. **They are metals with high m.p. and density** (Titanium is exceptional in being very light, zinc is exceptional in melting at a low temperature.)
2. **The elements and their compounds often act as catalysts** e.g. iron in the Haber process, vanadium(V) oxide in the Contact process, nickel in margarine manufacture.
3. **They form coloured compounds** e.g. potassium manganate(VII) is purple; potassium dichromate is orange; iron(II) sulphate is green; iron(III) nitrate is brown; copper(II) sulphate is blue.
4. **They can react with another element to form more than one compound** e.g. copper forms two oxides, copper(I) oxide, Cu_2O, and copper(II) oxide, CuO; iron forms two chlorides, iron(II) chloride, $FeCl_2$, and iron(III) chloride, $FeCl_3$.

Group 1

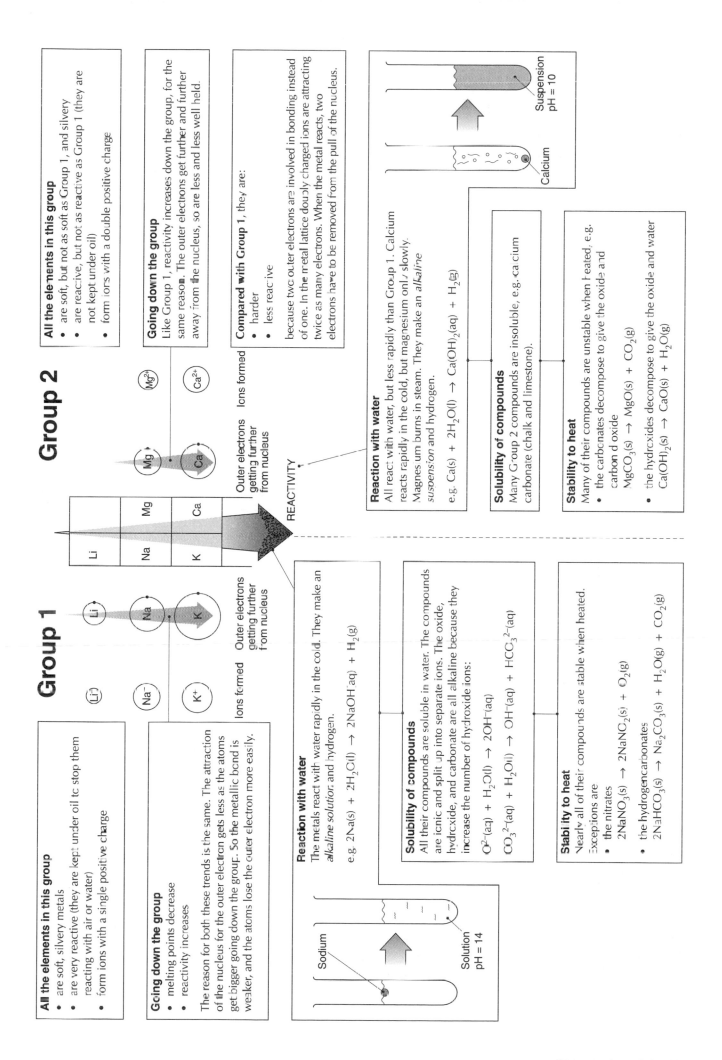

All the elements in this group

- are soft, silvery metals
- are very reactive (they are kept under oil to stop them reacting with air or water)
- form ions with a single positive charge

Going down the group

- melting points decrease
- reactivity increases

The reason for both these trends is the same. The attraction of the nucleus for the outer electron gets less as the atoms get bigger going down the group. So the metallic bond is weaker, and the atoms lose the outer electron more easily.

Li
Na
K

Outer electrons getting further from nucleus

Li⁻
Na⁻
K⁺

Ions formed

Reaction with water

The metals react with water rapidly in the cold. They make an *alkaline solution* and hydrogen.

e.g. $2Na(s) + 2H_2O(l) \rightarrow 2NaOH(aq) + H_2(g)$

Sodium

Solution pH = 14

Solubility of compounds

All their compounds are soluble in water. The compounds are ionic and split up into separate ions. The oxide, hydroxide, and carbonate are all alkaline because they increase the number of hydroxide ions:

$O^{2-}(aq) + H_2O(l) \rightarrow 2OH^-(aq)$

$CO_3^{2-}(aq) + H_2O(l) \rightarrow OH^-(aq) + HCO_3^{2-}(aq)$

Stability to heat

Nearly all of their compounds are stable when heated.
Exceptions are
- the nitrates
  $2NaNO_3(s) \rightarrow 2NaNO_2(s) + O_2(g)$
- the hydrogencarbonates
  $2NaHCO_3(s) \rightarrow Na_2CO_3(s) + H_2O(g) + CO_2(g)$

Group 2

All the elements in this group

- are soft, but not as soft as Group 1, and silvery
- are reactive, but not as reactive as Group 1 (they are not kept under oil)
- form ions with a double positive charge

Going down the group

Like Group 1, reactivity increases down the group, for the same reason. The outer electrons get further and further away from the nucleus, so are less and less well held.

Mg
Ca

Outer electrons getting further from nucleus

Mg²⁺
Ca²⁺

Ions formed

Compared with Group 1, they are:
- harder
- less reactive

because two outer electrons are involved in bonding instead of one. In the metal lattice doubly charged ions are attracting twice as many electrons. When the metal reacts, two electrons have to be removed from the pull of the nucleus.

REACTIVITY

Reaction with water

All react with water, but less rapidly than Group 1. Calcium reacts rapidly in the cold, but magnesium only slowly. Magnesium burns in steam. They make an *alkaline suspension* and hydrogen.

e.g. $Ca(s) + 2H_2O(l) \rightarrow Ca(OH)_2(aq) + H_2(g)$

Calcium

Suspension pH = 10

Solubility of compounds

Many Group 2 compounds are insoluble, e.g. calcium carbonate (chalk and limestone).

Stability to heat

Many of their compounds are unstable when heated, e.g.
- the carbonates decompose to give the oxide and carbon dioxide
  $MgCO_3(s) \rightarrow MgO(s) + CO_2(g)$
- the hydroxides decompose to give the oxide and water
  $Ca(OH)_2(s) \rightarrow CaO(s) + H_2O(g)$

Group 0, the noble gases

All the elements in this group

- are colourless gases
- are found in small amounts in the air
- are very, very unreactive (helium and neon never react, the others only very rarely and under special conditions)

The atoms are relatively small with full shells. They hold on to the electrons they already have strongly and have no room to gain any new electrons

Going down the group

- the elements get more dense
- the boiling points increase

density

0
He
Ne
Ar
Kr
Xe

Uses

Their uses depend on their physical properties or their unreactivity.

Helium
- in balloons (very low density)
- for deep diving (unreactive)
- used for super-cooling (low b.p.)

Neon
- in advertising signs (glows in electric field)

Argon
- to fill light bulbs (unreactive and cheap)
- in welding (unreactive)

Xenon
- in flash bulbs (unreactive). The unreactive gas prevents the filament of the bulb reacting when it flashes and gets very hot, so the bulb can be used many times.

Precipitate of silver halide

SILVER NITRATE

Halide solution

Group 7, the halogens

All the elements in this group

- are coloured non-metals; the colour darkens down the group.
- exist as **diatomic molecules** (molecules made of two atoms): F_2; Cl_2; Br_2; I_2
- have melting and boiling points which increase down the group
- are very reactive

Elements **Ions**

yellow gas $\textcircled{F}^-$

green gas Cl^-

red liquid Br^-

dark grey solid I^-

element reactivity

Going down the group

Reactivity *decreases* down the group, so each halogen displaces the ones below it from their compounds:

e.g. $Cl_2(g) + 2KBr(aq) \rightarrow Br_2(aq) + 2KCl(aq)$
$Cl_2(g) + 2KI(aq) \rightarrow I_2(aq) + 2KCl(aq)$
$Br_2(g) + 2KI(aq) \rightarrow I_2(aq) + 2KBr(aq)$

This trend in reactivity is opposite to that seen in Groups 1 and 2.

The reason for this is that these elements react by *gaining* electrons instead of losing them like metal atoms do. They usually react to form anions with a *single negative charge*. The smaller atoms at the top of the group attract electrons more strongly than the larger ones at the bottom. So the elements at the top of the group are more reactive than those below them.

Uses

Halogens are very **reactive oxidizing agents**. Solutions of chlorine in water and iodine in alcohol are used to kill bacteria.

Fluoride ions are added to toothpaste; a fluorine compound is the surface on non-stick pans.

Chlorine forms many covalent compounds with many important uses:

- CH_3CCl_3, 1,1,1-trichloroethane, is a very good solvent, used in dry cleaning
- CCl_2F_2, a chlorofluorocarbon (CFC) used as an aerosol propellant, now discontinued because of damage to the ozone layer.

Other compounds are used as anaesthetics, drugs, pesticides, and herbicides.

Reactions

All the elements in this group
- react with hydrogen forming hydrogen halides which are acidic in water
  e.g. $H_2(g) + Cl_2(g) \rightarrow 2HCl(g)$
  $HCl(g) + H_2O(l) \rightarrow H_3O^+(aq) + Cl^-(aq)$

- react with metals forming ionic metal halides
  e.g. $Na(s) + Cl_2(g) \rightarrow 2NaCl(s)$
  $2Al(s) + 3Cl_2(g) \rightarrow 2AlCl_3(s)$

Iron reacts similarly.

Precipitation reactions of halide ions

Halide ions in solution react with silver ions to form insoluble precipitates:

$Ag^+(aq) + Cl^-(aq) \rightarrow AgCl(s)$ white
$Ag^+(aq) + Br^-(aq) \rightarrow AgBr(s)$ cream
$Ag^+(aq) + I^-(aq) \rightarrow AgI(s)$ yellow

These solids darken in light and form the basis of photographic film.

Sulphur chemistry

SULPHURIC ACID, H$_2$SO$_4$

Sulphuric acid is probably the most important industrial chemical made because it is used in so many processes. It is made from sulphur in the **Contact process.**

The Contact process

- **sulphur is melted and burnt in air**

$$S(l) + O_2(g) \rightarrow SO_2(g)$$

This is an exothermic combination reaction. The sulphur burns easily and the reaction goes to completion.

- **the sulphur dioxide is oxidized to sulphur trioxide**

$$2SO_2(g) + O_2(g) \underset{V_2O_5}{\overset{450°C}{\rightleftharpoons}} 2SO_3(g)$$

This is also an exothermic reaction, but it is reversible.

High rate would be produced by high pressure, high temperature, and a catalyst.

High yield would be produced by high pressure, but low temperature.

The actual conditions used are a compromise.

High pressure is too expensive so a pressure just above atmospheric is used. At 450°C, 97% conversion is achieved.

- **the sulphur trioxide is hydrated with water**

$$H_2O(l) + SO_3(g) \rightarrow H_2SO_4(l)$$

This is another very exothermic reaction. Just adding the sulphur trioxide gas to water would make the water boil, forming a stable acid mist. To stop this, the sulphur trioxide is bubbled into 98% sulphuric acid, 2% water. This mixture has a much higher boiling point. The water reacts making even more acid without making a mist. So no acid vapours escape from the plant.

Uses

Fertilizer
30%

30%

14%

11%

3%

5%

Other uses

Plastics

Fibres

Paint

Detergents

SULPHUR

Sulphur is a non-metallic element in Group 6. Its properties are typical of a non-metal.

- **Physical:** brittle, insulating, molecular solid, S$_8$.
- **Chemical:** forms an anion, S^{2-}, and an acidic oxide, SO$_2$.

$$SO_2(g) + H_2O(l) \rightarrow H_2SO_3(aq) \text{ sulphurous acid.}$$

SULPHUR DIOXIDE, SO$_2$

Sulphur dioxide is a covalent, molecular substance, with weak forces between the molecules. It is a dense gas.

It can be oxidized or reduced depending on what it reacts with:

- hydrogen sulphide reduces it to sulphur

$$2H_2S(g) + SO_2(g) \rightarrow 3S(s) + 2H_2O(l)$$

- potassium dichromate oxidizes it in solution; the colour of the dichromate changes from orange to green.

$$K_2Cr_2O_7(aq) + 3H_2SO_3(aq) + H_2SO_4(aq)$$
$$\underset{orange}{} \rightarrow \underset{green}{Cr_2(SO_4)_3(aq)} + K_2SO_4(aq) + 4H_2O(l)$$

Reactions of concentrated sulphuric acid

- **As a strong acid**

$$H_2SO_4(l) + 2H_2O(l) \rightarrow 2H_3O^+(aq) + SO_4^{2-}(aq)$$

- **As an oxidizing agent:** it will oxidize both metals and non-metals.

$$Cu(s) + 2H_2SO_4(l) \rightarrow CuSO_4(aq) + SO_2(g) + 2H_2O(l)$$
$$C(s) + 2H_2SO_4(l) \rightarrow CO_2(g) + 2SO_2(g) + 2H_2O(l)$$

- **As a dehydrating agent:** it will remove water or the elements of water. It makes hydrated copper(II) sulphate into the anhydrous form

$$CuSO_4.5H_2O(s) \rightarrow CuSO_4(s)$$

It dehydrates sucrose

$$C_{12}H_{22}O_{11}(s) \rightarrow 12C(s)$$

Important ideas about chemical change 1: making new substances

In a chemical reaction

The starting substances – the reactants – react to give new, different substances – the products. The changes which take place in the reaction are usually written as an equation.

reactants → products

means 'reacts to give' (sometimes the reaction conditions are written over the arrow)

Word equations

These express in words what is reacting with what to make the products.

e.g. magnesium metal and gaseous oxygen → solid magnesium oxide

Symbol equations

These express in terms of chemical symbols the formulas of the substances taking part in the reaction. They are balanced in the sense that there are the same number of each atom on each side of the arrow.

e.g. $2Mg + O_2 \rightarrow 2MgO$

State symbols

These tell us even more about the reaction. After the formula of each substance there are state symbols which describe the state that the substance is in.

So using (s) for solid, (l) for liquid, (aq) for a solution in water, and (g) for gas, the equation becomes:
$2Mg(s) + O_2(g) \rightarrow 2MgO(s)$

DIFFERENT KINDS OF CHEMICAL CHANGE

Decomposition

Here a single substance is broken down into two or more simpler substances.

e.g. most metal carbonates decompose to give the oxide and carbon dioxide:
$CuCO_3(s) \rightarrow CuO(s) + CO_2(g)$

Remember, substances which cannot be decomposed because they contain only one kind of atom are called elements.

Heat in, so endothermic

Combination

Here two substances (usually elements) react together to make a single new compound.

e.g. metal and non-metal: aluminium and iodine react to give aluminium iodide:
$2Al(s) + 3I_2(s) \rightarrow 2AlI_3(s)$

Sometimes two elements combine together in more than one way.

e.g. carbon can react with oxygen to make two different oxides: carbon monoxide, CO, and carbon dioxide, CO_2. Which one forms depends on how much oxygen is available for the reaction.

Heat out, so exothermic

Electrolysis

Electrolysis is a form of decomposition. It is used to extract very reactive metals from their ores. The ores of these reactive metals are very stable and cannot be decomposed in any other way.

The metal ions go to the cathode and the non-metal ions go to the anode.

e.g. in molten sodium chloride, the reaction is:
$2NaCl(l)$
$\rightarrow 2Na(l) + Cl_2(g)$

Energy in, so endothermic

Displacement

These are reactions in which one element takes the place of another. Both metals and non-metals can do displacements.

More reactive metals can displace less reactive ones from their solutions:
$Fe(s) + CuSO_4(aq) \rightarrow FeSO_4(aq) + Cu(s)$

Reactive metals can displace hydrogen from acids:
$Mg(s) + H_2SO_4(aq) \rightarrow MgSO_4(aq) + H_2(g)$

Very reactive metals can even displace hydrogen from water:
$2Na(s) + 2H_2O(l) \rightarrow 2NaOH(aq) + H_2(g)$

More reactive non-metals can displace less reactive ones from their solutions:
$Cl_2(g) + 2KI(aq) \rightarrow I_2(s) + 2KCl(aq)$

Displacement reactions are a good way of comparing the reactivity of elements.

Important ideas about chemical change 2: energy

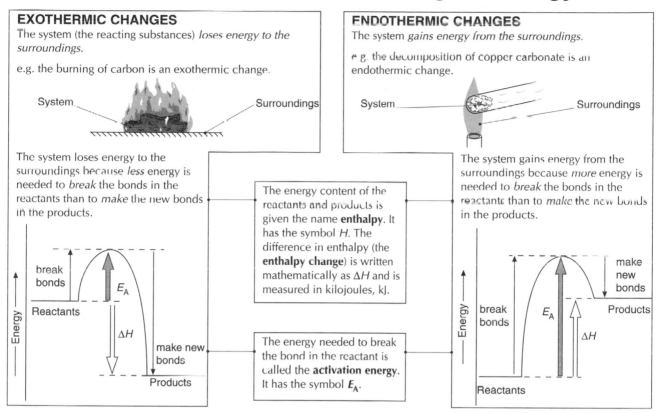

EXOTHERMIC CHANGES
The system (the reacting substances) *loses energy to the surroundings.*

e.g. the burning of carbon is an exothermic change.

System ———— ———— Surroundings

The system loses energy to the surroundings because *less* energy is needed to *break* the bonds in the reactants than to *make* the new bonds in the products.

break bonds

E_A

Reactants

ΔH

make new bonds

Products

Energy →

ENDOTHERMIC CHANGES
The system *gains energy from the surroundings.*

e.g. the decomposition of copper carbonate is an endothermic change.

System ———— ———— Surroundings

The system gains energy from the surroundings because *more* energy is needed to *break* the bonds in the reactants than to *make* the new bonds in the products.

make new bonds

break bonds

E_A

ΔH

Products

Reactants

Energy →

The energy content of the reactants and products is given the name **enthalpy**. It has the symbol H. The difference in enthalpy (the **enthalpy change**) is written mathematically as ΔH and is measured in kilojoules, kJ.

The energy needed to break the bond in the reactant is called the **activation energy**. It has the symbol E_A.

FUELS
Fuels are substances which react exothermically with air making safe products. They combust (burn) easily, so storing them can be a problem.

Hydrogen reacts exothermically and the product (water) is totally safe. But hydrogen is difficult and dangerous to store.

$2H_2(g) + O_2(g) \longrightarrow 2H_2O(g)$

Bonds broken	Bonds made	Total energy change
2 H–H = 2 × 436	4 H–O = 4 × 463	
1 O=O = 496		
1368 kJ	1852 kJ	1852 – 1368 = 484 kJ

Methane (natural or North Sea gas) reacts exothermically. The products are carbon dioxide and water. These are safe, although carbon dioxide contributes to global warming.

$CH_4(g) + 2O_2(g) \longrightarrow CO_2(g) + 2H_2O(g)$

Bonds broken	Bonds made	Total energy change
4 C–H = 4 × 412	2 C=O = 2 × 743	
2 O=O = 2 × 496	4 H–O = 4 × 463	3338 – 2640 = 698 kJ
2640	3338	

Comparing fuels
The most efficient fuel is one that gives out most energy per gram.

4 g of hydrogen produce 484 kJ, so 1 g → 121 kJ
16 g of methane produce 698 kJ, so 1 g → 43.6 kJ

Hydrogen in the more efficient fuel.

FIRE
Fire triangle
For a fire to burn, fuel, air, and heat are needed. The heat provides the **activation energy**, E_A, to break existing bonds.

These three requirements are shown in the fire triangle.

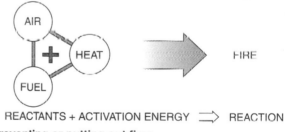

AIR

HEAT

FUEL

FIRE

REACTANTS + ACTIVATION ENERGY $\Longrightarrow$ REACTION

Preventing or putting out fires
A fire can be prevented by removing any one of the components of the fire triangle:
* remove the air (fire blankets, foam)
* remove the fuel (turn off fuel, use fire breaks in forests)
* remove heat (spray with water)

INCOMPLETE COMBUSTION
In the open air, a fuel reacts with oxygen until it is used up. The fuel is **completely combusted**.

If the supply of air is limited, the oxygen in the air may be used up before the fuel. The fuel is incompletely combusted.

If a hydrocarbon is **incompletely combusted**, the carbon in it may be:
* only partially oxidized, forming carbon monoxide
* unburnt, forming soot (as in a yellow Bunsen flame)

In car engines, turbos pump more air into the engine. This improves efficiency because it helps the fuel to combust completely.

Important ideas about chemical change 3: reactivity

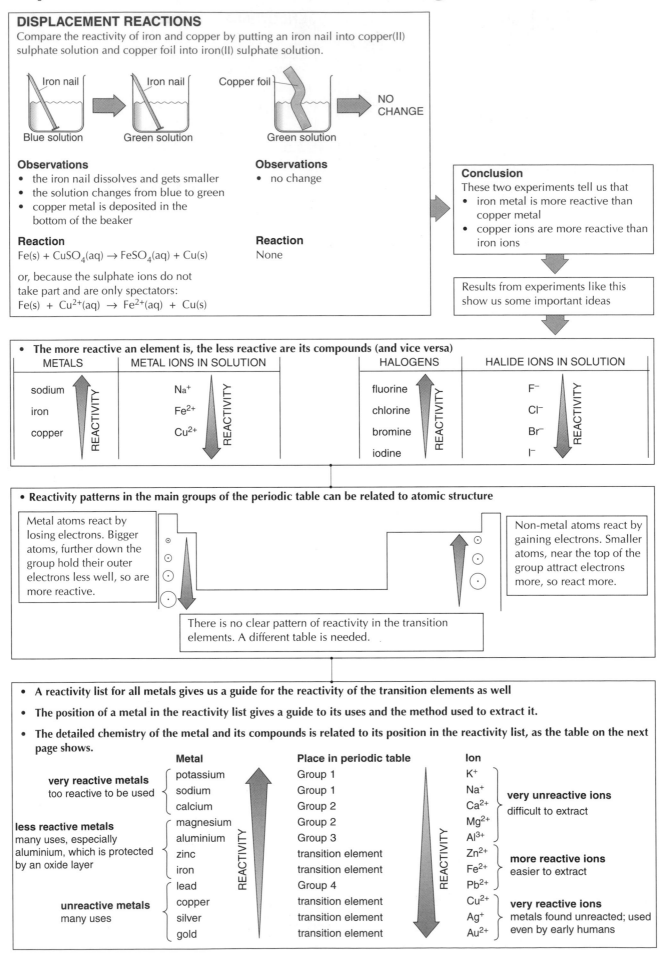

DISPLACEMENT REACTIONS

Compare the reactivity of iron and copper by putting an iron nail into copper(II) sulphate solution and copper foil into iron(II) sulphate solution.

Iron nail — Blue solution → Iron nail — Green solution

Copper foil — Green solution → NO CHANGE

Observations
- the iron nail dissolves and gets smaller
- the solution changes from blue to green
- copper metal is deposited in the bottom of the beaker

Reaction

$Fe(s) + CuSO_4(aq) \rightarrow FeSO_4(aq) + Cu(s)$

or, because the sulphate ions do not take part and are only spectators:

$Fe(s) + Cu^{2+}(aq) \rightarrow Fe^{2+}(aq) + Cu(s)$

Observations
- no change

Reaction

None

Conclusion

These two experiments tell us that
- iron metal is more reactive than copper metal
- copper ions are more reactive than iron ions

Results from experiments like this show us some important ideas

---

- **The more reactive an element is, the less reactive are its compounds (and vice versa)**

METALS	METAL IONS IN SOLUTION		HALOGENS	HALIDE IONS IN SOLUTION
sodium	Na^+		fluorine	F^-
iron	Fe^{2+}		chlorine	Cl^-
copper	Cu^{2+}		bromine	Br^-
			iodine	I^-

REACTIVITY (for each column)

- **Reactivity patterns in the main groups of the periodic table can be related to atomic structure**

Metal atoms react by losing electrons. Bigger atoms, further down the group hold their outer electrons less well, so are more reactive.

Non-metal atoms react by gaining electrons. Smaller atoms, near the top of the group attract electrons more, so react more.

There is no clear pattern of reactivity in the transition elements. A different table is needed.

- **A reactivity list for all metals gives us a guide for the reactivity of the transition elements as well**
- **The position of a metal in the reactivity list gives a guide to its uses and the method used to extract it.**
- **The detailed chemistry of the metal and its compounds is related to its position in the reactivity list, as the table on the next page shows.**

	Metal	Place in periodic table	Ion	
very reactive metals too reactive to be used	potassium	Group 1	K^+	**very unreactive ions** difficult to extract
	sodium	Group 1	Na^+	
	calcium	Group 2	Ca^{2+}	
less reactive metals many uses, especially aluminium, which is protected by an oxide layer	magnesium	Group 2	Mg^{2+}	
	aluminium	Group 3	Al^{3+}	
	zinc	transition element	Zn^{2+}	**more reactive ions** easier to extract
	iron	transition element	Fe^{2+}	
	lead	Group 4	Pb^{2+}	
unreactive metals many uses	copper	transition element	Cu^{2+}	**very reactive ions** metals found unreacted; used even by early humans
	silver	transition element	Ag^+	
	gold	transition element	Au^{2+}	

REACTIVITY

Metal reactivity and uses

METAL	REACTION WITH AIR	REACTION WITH WATER	REACTION WITH DILUTE ACID	EXTRACTION	USES	METAL
potassium		react quickly in cold water				potassium
sodium					**street lamps:** sodium and its compounds give orange light when heated	sodium
calcium	burn to form oxide		all these metals displace hydrogen from dilute acids; reactivity increases up the list	**electrolysis** of molten compound	**steel production:** calcium reacts with and removes oxygen	calcium
magnesium		burns in steam			**alloy with aluminium:** makes aluminium stronger **flares:** burns with a very bright light	magnesium
aluminium		oxide layer stops reaction			**major structural metal:** strong, but light; protected by oxide layer	aluminium
zinc		reacts in steam		reaction with carbon in blast furnace	**alloy with copper:** brass **sacrificial protection of steel:** galvanizing	zinc
iron		reacts reversibly in steam			**major structural metal:** cheap to extract and strong when alloyed with carbon, but it rusts	iron
hydrogen	react slowly when heated					hydrogen
copper		no reaction	metals below hydrogen never react with acids to displace hydrogen	decomposition by heat alone	**pipes:** ductile but unreactive **coinage:** colour, unreactive **electrical cables:** good conductor	copper
gold					**jewellery:** rare, attractive colour **electrical contacts:** good conductor, unreactive	gold

Notes
Hydrogen is in the list because many metal reactions involve the metal displacing hydrogen from water or acids.

Group 1 metals are kept under oil to stop reaction with air. Other metals form oxide layers which slow down further reaction.

The reactivity of aluminium appears *anomalous* or *unexpected*. It is a very reactive metal, but it appears to be unreactive because the oxide layer protects it.

Unreactive metals do not react to produce hydrogen. But some of them react with acids in other ways. For example, copper reacts with dilute nitric acid. In this reaction nitric acid is not acting as an acid (through the hydrogen ion) but as an oxidizing agent (through the nitrate ion).

The reactive metals have such unreactive compounds that they can only be decomposed by electrolysis.
The unreactive metals at the bottom of the list have such reactive compounds that the metals were discovered and used in prehistoric times.

Metal uses depend on a number of factors:
- **abundance of the metal:** aluminium is more abundant than iron; gold is very rare.
- **ease of extraction:** gold and iron are easy to extract; aluminium is expensive to extract because so much electricity is needed.
- **suitable physical and chemical properties:** these are listed above.

Water

BONDING AND STRUCTURE

slightly positive — slightly negative

slightly positive

The shared pairs in the covalent bonds are pulled nearer to the oxygen than the hydrogen.

This makes the hydrogen ends of the molecule a bit positive and the oxygen part a bit negative. This means the molecule is **polar**.

TESTS FOR WATER

Water can be recognized by
- its physical properties
It freezes at 0°C and boils at 100°C.
- its solvent properties
It dissolves anhydrous white copper sulphate turning it blue:
$$CuSO_4(s) \rightarrow CuSO_4.5H_2O(aq)$$
white   blue

It dissolves anhydrous blue cobalt chloride turning it pink.

$$CoCl_2(s) \rightarrow CoCl_2.7H_2O(aq)$$
blue   pink

SOLVENT PROPERTIES

The charged parts of each water molecule attract other molecules and ions making water a very good solvent.

Ionic solids (like salt) and **covalent solids** (like sugar) all get *more* soluble as the temperature increases.

Gases (like oxygen, carbon dioxide, and hydrogen chloride) get *less* soluble as the temperature increases. The oxygen dissolved in water is essential to aquatic life.

PHYSICAL PROPERTIES

hydrogen bond

The charged parts of each molecule attract oppositely charged parts on nearby molecules strongly. These forces are called **hydrogen bonds**. They hold water molecules together tightly, so water is a liquid not a gas at room temperature.

PURIFICATION

Because water is such a good solvent, it is difficult to find it pure. **Rainwater** contains dissolved gases. It dissolves gases like carbon dioxide (which occurs in the atmosphere naturally), and sulphur dioxide and nitrogen dioxide (which are in the atmosphere as a result of human activities).

AMMONIA IN SOLUTION

Ammonia gas is very soluble in water. It *takes a proton from water* (unlike the hydrogen halides which give a proton to water). The resulting solution is alkaline, because it contains hydroxide ions, OH⁻.

$$NH_3(g) + H_2O(l) \rightarrow NH_4^+(aq) + OH^-(aq)$$

HYDROGEN CHLORIDE IN SOLUTION

Hydrogen chloride and the other hydrogen halides are very soluble in water. They react with the water making *strongly acidic* solutions because they *give a proton to water* making the hydroxonium ion, H_3O^+. This is usually written more simply as H^+. So the reaction equation is

$$H_2O(l) + HCl(g) \rightarrow H_3O^+(aq) + Cl^-(aq)$$

$$or \quad HCl(g) \xrightarrow{H_2O} H^+(aq) + Cl^-(aq)$$

The resulting solution is called hydrochloric acid.

SALTS IN SOLUTION

Many ionic solids dissolve in water. Sea water is a source of these salts and the elements in them.

Sea water contains sodium chloride and small amounts of other metal halides. Bromine is extracted from the magnesium bromide in sea water by displacement with chlorine.

$$Cl_2(g) + MgBr_2(aq) \rightarrow MgCl_2(aq) + Br_2(aq)$$

CARBON DIOXIDE IN SOLUTION

Carbon dioxide dissolves in water forming **carbonic acid**.

Rain is very dilute carbonic acid; soda water is a more concentrated solution.

$$H_2O(l) + CO_2(g) \rightarrow H_2CO_3(aq)$$

The water cycle

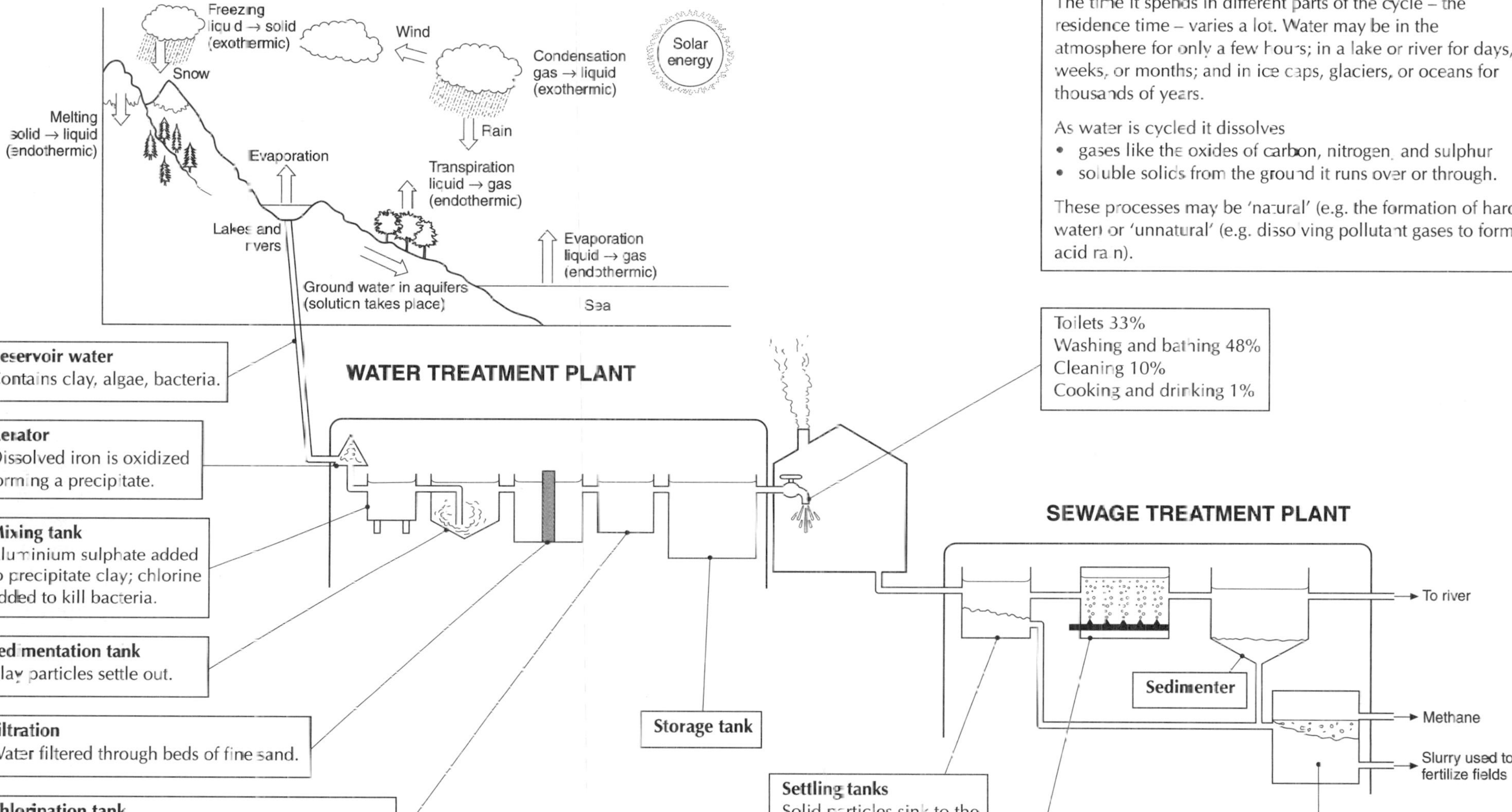

WATER AS A RESOURCE

Because water is cycled quickly it is a renewable resource. The time it spends in different parts of the cycle – the residence time – varies a lot. Water may be in the atmosphere for only a few hours; in a lake or river for days, weeks, or months; and in ice caps, glaciers, or oceans for thousands of years.

As water is cycled it dissolves
- gases like the oxides of carbon, nitrogen, and sulphur
- soluble solids from the ground it runs over or through.

These processes may be 'natural' (e.g. the formation of hard water) or 'unnatural' (e.g. dissolving pollutant gases to form acid rain).

Solar energy

Freezing
liquid → solid
(exothermic)

Snow

Melting
solid → liquid
(endothermic)

Wind

Condensation
gas → liquid
(exothermic)

Rain

Transpiration
liquid → gas
(endothermic)

Evaporation

Lakes and rivers

Ground water in aquifers
(solution takes place)

Evaporation
liquid → gas
(endothermic)

Sea

Toilets 33%
Washing and bathing 48%
Cleaning 10%
Cooking and drinking 1%

WATER TREATMENT PLANT

Reservoir water
Contains clay, algae, bacteria.

Aerator
Dissolved iron is oxidized forming a precipitate.

Mixing tank
Aluminium sulphate added to precipitate clay; chlorine added to kill bacteria.

Sedimentation tank
Clay particles settle out.

Filtration
Water filtered through beds of fine sand.

Chlorination tank
More chlorine added to kill all bacteria. Chlorine is a strong oxidizing agent and reacts with bacteria.

Storage tank

Water treatment plants cannot remove soluble impurities such as nitrates. So the water running off fields into rivers is monitored to check that leached nitrates stay below safe limits.

SEWAGE TREATMENT PLANT

To river

Methane

Slurry used to fertilize fields

Sedimenter

Settling tanks
Solid particles sink to the bottom making sludge.

Aeration tank
Air is blown through water. Aerobic bacteria contain enzymes which catalyse the decomposition of harmful substances in the water.

Sludge digester
Anaerobic bacteria contain enzymes which catalyse decomposition of harmful substances. Methane gas and fertilizer are produced.

Solubility and solutions in water

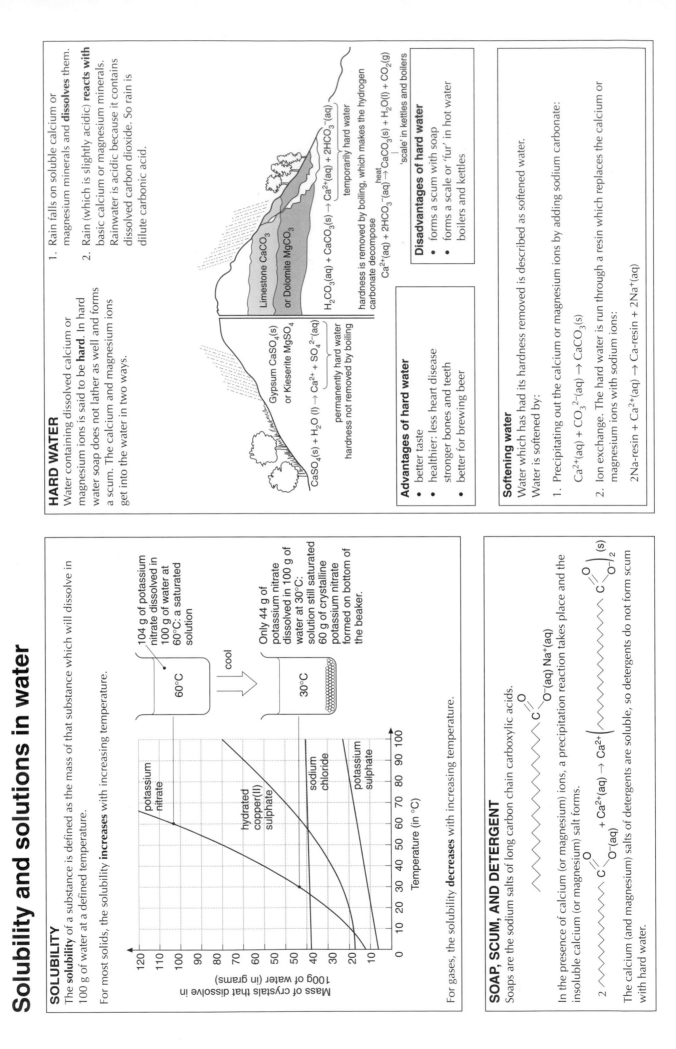

SOLUBILITY

The **solubility** of a substance is defined as the mass of that substance which will dissolve in 100 g of water at a defined temperature.

For most solids, the solubility **increases** with increasing temperature.

Graph: Mass of crystals that dissolve in 100g of water (in grams) vs Temperature (in °C). Curves for potassium nitrate, hydrated copper(II) sulphate, sodium chloride, potassium sulphate.

104 g of potassium nitrate dissolved in 100 g of water at 60°C: a saturated solution

60°C → cool → 30°C

Only 44 g of potassium nitrate dissolved in 100 g of water at 30°C: solution still saturated 60 g of crystalline potassium nitrate formed on bottom of the beaker.

For gases, the solubility **decreases** with increasing temperature.

SOAP, SCUM, AND DETERGENT

Soaps are the sodium salts of long carbon chain carboxylic acids.

(structural formula) $\text{C}{=}\text{O}\ \text{O}^-(aq)\ Na^+(aq)$

In the presence of calcium (or magnesium) ions, a precipitation reaction takes place and the insoluble calcium (or magnesium) salt forms.

$2\ (\text{chain})\ \text{C}{=}\text{O}\ \text{O}^-(aq) + Ca^{2+}(aq) \rightarrow Ca^{2+}\left((\text{chain})\ \text{C}{=}\text{O}\ \text{O}^-\right)_2 (s)$

The calcium (and magnesium) salts of detergents are soluble, so detergents do not form scum with hard water.

HARD WATER

Water containing dissolved calcium or magnesium ions is said to be **hard**. In hard water soap does not lather as well and forms a scum. The calcium and magnesium ions get into the water in two ways.

1. Rain falls on soluble calcium or magnesium minerals and **dissolves** them.
2. Rain (which is slightly acidic) **reacts with** basic calcium or magnesium minerals. Rainwater is acidic because it contains dissolved carbon dioxide. So rain is dilute carbonic acid.

Limestone $CaCO_3$ or Dolomite $MgCO_3$

$H_2CO_3(aq) + CaCO_3(s) \rightarrow Ca^{2+}(aq) + 2HCO_3^-(aq)$
temporarily hard water

Gypsum $CaSO_4(s)$ or Kieserite $MgSO_4$

$CaSO_4(s) + H_2O(l) \rightarrow Ca^{2+} + SO_4^{2-}(aq)$
permanently hard water hardness not removed by boiling

hardness is removed by boiling, which makes the hydrogen carbonate decompose
$Ca^{2+}(aq) + 2HCO_3^-(aq) \xrightarrow{heat} CaCO_3(s) + H_2O(l) + CO_2(g)$
'scale' in kettles and boilers

Disadvantages of hard water
• forms a scum with soap
• forms a scale or 'fur' in hot water boilers and kettles

Advantages of hard water
• better taste
• healthier: less heart disease
• stronger bones and teeth
• better for brewing beer

Softening water

Water which has had its hardness removed is described as softened water. Water is softened by:

1. Precipitating out the calcium or magnesium ions by adding sodium carbonate:
$Ca^{2+}(aq) + CO_3^{2-}(aq) \rightarrow CaCO_3(s)$

2. Ion exchange. The hard water is run through a resin which replaces the calcium or magnesium ions with sodium ions:
$2Na\text{-resin} + Ca^{2+}(aq) \rightarrow Ca\text{-resin} + 2Na^+(aq)$

Acids and bases

ACIDS

Acid properties

- sour taste: citric acid in lemons and grapefruits; ethanoic acid in vinegar (never taste acids in the lab)
- change the colour of indicators: litmus and universal indicators go red
- react with water to make hydrogen ions, H_3O^+, or more simply, $H^+(aq)$
  $$H_2O(l) + HNO_3(g) \rightarrow H^+(aq) + NO^{3-}(aq)$$
- react with reactive metals which displace the hydrogen in the acid
  $$Mg(s) + H_2SO_4(aq) \rightarrow MgSO_4(aq) + H_2(g)$$
- react with metal oxides, metal hydroxides, and metal carbonates to make salts; when this happens the acid is *neutralized*.

Strong and weak acids

Some acids react completely with water forming hydrogen ions:

$$H_2O(l) + HCl(g) \rightarrow H^+(aq) + Cl^-(aq)$$

The solution made conducts strongly because it contains so many ions. These acids are called strong acids.

Other acids react incompletely with water, and only make a few hydrogen ions:

$$H_2O(l) + CH_3COOH(l) \rightleftharpoons H^+(aq) + CH_3COO^-(aq)$$
$$99.9\% \qquad\qquad 0.1\% \qquad 0.1\%$$

Because there are so few ions present in the solution, it conducts weakly. Acids like this are called weak acids.

So
- **strong acids are fully ionized** in water
- **weak acids are only partially ionized** in water.

Definition of an acid

An acid is a substance which contains hydrogen which can be displaced by a metal and which reacts with water to make hydrogen ions as the only positive ions.

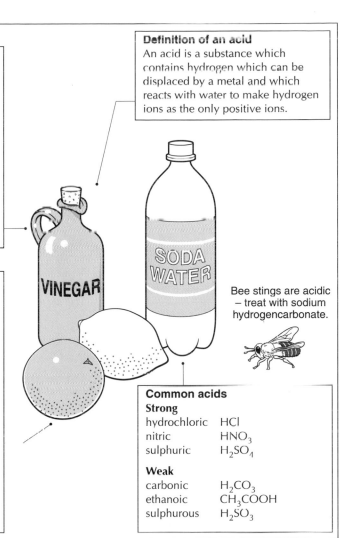

Bee stings are acidic – treat with sodium hydrogencarbonate.

Common acids

Strong

hydrochloric	HCl
nitric	HNO_3
sulphuric	H_2SO_4

Weak

carbonic	H_2CO_3
ethanoic	CH_3COOH
sulphurous	H_2SO_3

BASES

Bases and alkalis

Substances which neutralize acids are called **bases**.

Bases are the oxides, hydroxides, or carbonates of metals.

Most bases are insoluble, but some dissolve in water.

Soluble bases are called **alkalis**.

Strong and weak alkalis

Some alkalis split up completely into ions:
$$NaOH(s) + H_2O(l) \rightarrow$$
$$Na^+(aq) + OH^-(aq)$$

They are called strong alkalis.

Other alkalis react incompletely with water making only a few hydroxide ions:
$$NH_3(aq) + H_2O(l) \rightleftharpoons$$
$$NH_4^+(aq) + OH^-(aq)$$

They are called weak alkalis.

Common alkalis

Group 1 oxides, hydroxides, and carbonates: $NaOH$; K_2O; Na_2CO_3

Calcium hydroxide (lime water): $Ca(OH)_2$

Ammonia solution (aqueous ammonia): $NH_3(aq)$

Wasp stings are alkaline – treat with lemon juice or vinegar.

Alkaline solutions

The solution made from a soluble base:
- has a soapy, slippery feel – most soaps are alkaline
- changes the colour of indicators – litmus and universal go blue.

Neutralization

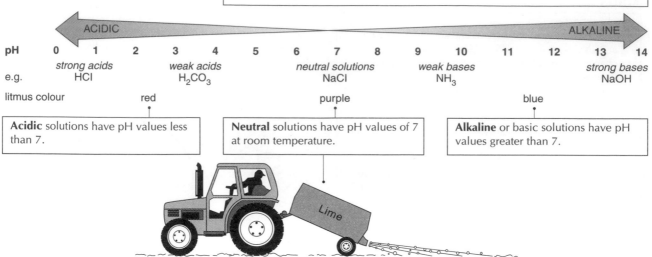

pH	0	1	2	3	4	5	6	7	8	9	10	11	12	13	14
e.g.	strong acids HCl			weak acids H_2CO_3				neutral solutions NaCl		weak bases NH_3					strong bases NaOH

litmus colour   red   purple   blue

Acidic solutions have pH values less than 7.

Neutral solutions have pH values of 7 at room temperature.

Alkaline or basic solutions have pH values greater than 7.

NEUTRALIZATION

When an acid and a base react together they **neutralize** each other. The acid loses its acidic properties and the base loses its basic properties. The products of a neutralization reaction are a salt and water.

acid + base → salt + water (+ carbon dioxide if the base is a carbonate)

e.g. $NaOH + HCl \rightarrow NaCl + H_2O$
$MgO + H_2SO_4 \rightarrow MgSO_4 + H_2O$
$CaCO_3 + 2HNO_3 \rightarrow Ca(NO_3)_2 + H_2O + CO_2$

During these reactions:

- the hydrogen ions from the acid react with the hydroxide ions from the base to make water molecules

 $H^+(aq) + OH^-(aq) \rightarrow H_2O(l)$

- all the other ions are spectators and just stay in solution

- heat is produced because all neutralizations are exothermic

- the pH changes

- the colour of an indicator changes.

USING NEUTRALIZATIONS

There are a number of important examples of neutralization.

1. Adding lime (calcium hydroxide) to soil

Plants need nitrogen, phosphorus, and potassium compounds from the soil to grow well. Most plants take up these elements better when the soil is alkaline. Lime neutralizes acids in the soil making it alkaline.

2. Reducing acid rain and its effects

The coal burnt in power stations contains sulphur as an impurity. When the coal is burnt, the sulphur is burnt too. This produces sulphur dioxide, which dissolves in rain forming acid rain. By passing the burnt gases through lime, they are neutralized. The product, calcium sulphate, is used as plaster.

Many lakes have become acidic due to acid rain. All the fish in them die. Spraying powdered lime on to the lake neutralizes the water so that fish can once again survive.

3. Neutralizing stomach acids

The human stomach contains hydrochloric acid so that the enzyme pepsin can begin digesting protein. Sometimes too much acid is made and begins to attack the stomach wall causing pain. The extra acid can be neutralized by taking an 'ant-acid'. These always contain a base such as sodium or magnesium carbonate.

ACIDS WITH TWO HYDROGENS

Some acids like sulphuric, H_2SO_4, and carbonic, H_2CO_3, have two hydrogens in them. These acids can be

- **completely neutralized** when both hydrogens are replaced, e.g.
 $H_2CO_3 + 2NaOH \rightarrow Na_2CO_3 + 2H_2O$
 complete neutralization

- or **partly neutralized** when only one hydrogen is replaced, e.g.
 $H_2CO_3 + 1NaOH \rightarrow NaHCO_3 + 1H_2O$
 partial neutralization

The salt made by partial neutralization is called an **acid salt**. In this example the salt is called **sodium hydrogencarbonate**.

SALTS

Salts are substances in which the hydrogen of an acid has been replaced by a metal, e.g. NaCl or $MgSO_4$.

Each acid can produce a family of salts.

e.g.

    hydrochloric acid produces **chlorides**
    nitric acid produces **nitrates**
    sulphuric acid produces **sulphates**
    ethanoic acid produces **ethanoates**

The first part of the salt, the metal part, is determined by the base used to react with the acid. So each base also produces a family of salts.

e.g.

    sodium bases like sodium hydroxide produce **sodium salts**
    magnesium bases like magnesium oxide produce **magnesium salts**

Crystallization and precipitation

CRYSTALLIZATION

If the water in a solution is lost by evaporation, the ions that are left will clump together. The positive cations will attract the negative anions, and the ions will arrange themselves so that oppositely charged ions are next to each other. A lattice will slowly form. This process is called **crystallization**.

e.g. $Na^+(aq) + Cl^-(aq) \rightarrow NaCl(s)$

If the process happens slowly, big crystals have time to grow. If the process happens quickly, lots of small crystals form.

SLOW

Cold, saturated solution

Warm, saturated solution

PRECIPITATION

If two solutions containing oppositely charged ions which attract each other very strongly are added together, the ions will immediately clump together. Because this happens very quickly, there is not time for an ordered lattice to form. So instead of crystals, a fine suspension is seen. This is called a **precipitate**.

Precipitation is the reaction between two solutions making an *insoluble* product.

e.g.
$$CaCl_2(aq) + Na_2CO_3(aq) \rightarrow CaCO_3(s) + 2NaCl(aq)$$

or much better
$$Ca^{2+}(aq) + CO_3^{2-}(aq) \rightarrow CaCO_3(s)$$

because the other ions are spectators.

FAST

Opaque suspension

Transparent solutions

Using precipitation reactions

1. Making insoluble salts

Any insoluble salt can be made simply by adding together two solutions containing between them the ions in the salt.

So to make lead iodide, add any lead solution to any iodide solution, filter off and wash the precipitate, then dry it.

2. Testing for the presence of certain ions

Precipitation reactions can be used to test for both anions and cations.

Anion tests

Anions like the halides and sulphates can be tested for using precipitation reactions. For example, the presence of chloride ions in solution can be detected by adding silver ions. Silver ions attract chloride ions so strongly that if there are any chloride ions present a white precipitate of silver chloride will form.

(Equally silver ions can be tested for using chloride ions.)

A systematic scheme for testing for anions is shown on page 67.

Cation tests

Many metals form insoluble hydroxides. So if a solution of sodium hydroxide is added to a solution of the metal, a precipitate of the hydroxide will be seen. Hydroxide precipitates vary in colour, and some redissolve when excess sodium hydroxide is added.

A systematic scheme for testing for cations is shown on page 66.

Reduction and oxidation

REDUCTION

Reduction is the removal of oxygen. The word was originally used to describe the process of changing a metal ore (often an oxide) into the metal.

e.g. $2CuO(s) + C(s) \rightarrow 2Cu(s) + CO_2(g)$

The metal oxide above would not have been reduced without the carbon, so the carbon is called a **reducing agent**.

Notice that in acting as a reducing agent, the carbon is itself oxidized. This means that the copper oxide was acting as an **oxidizing agent**.

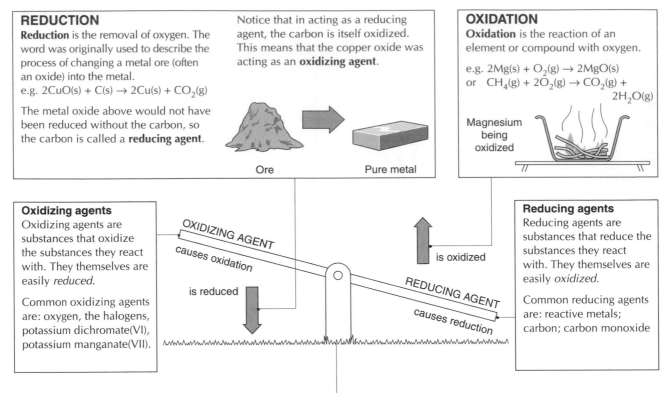

Ore   Pure metal

OXIDATION

Oxidation is the reaction of an element or compound with oxygen.

e.g. $2Mg(s) + O_2(g) \rightarrow 2MgO(s)$
or $CH_4(g) + 2O_2(g) \rightarrow CO_2(g) + 2H_2O(g)$

Magnesium being oxidized

Oxidizing agents

Oxidizing agents are substances that oxidize the substances they react with. They themselves are easily *reduced*.

Common oxidizing agents are: oxygen, the halogens, potassium dichromate(VI), potassium manganate(VII).

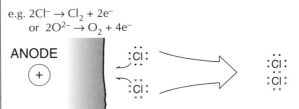

OXIDIZING AGENT
causes oxidation
is reduced

is oxidized

REDUCING AGENT
causes reduction

Reducing agents

Reducing agents are substances that reduce the substances they react with. They themselves are easily *oxidized*.

Common reducing agents are: reactive metals; carbon; carbon monoxide

REDOX

Reduction and oxidation take place together. All oxidation reactions involve the reduction of something else, so whether the process is called oxidation or reduction depends on which reactant is being referred to. So the whole process is often called **redox**.

Electron transfer

Magnesium atoms and oxygen molecules react to form magnesium oxide, an ionic solid. In this solid there are magnesium ions, Mg^{2+}, and oxide ions, O^{2-}.

So the magnesium (which is oxidized) loses electrons as the atoms become positive cations:
$Mg \rightarrow Mg^{2+} + 2e^-$

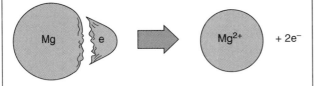

while the oxygen (which is reduced) gains electrons as the molecules become negative anions:
$O_2 + 4e^- \rightarrow 2O^{2-}$

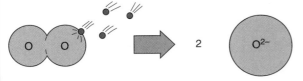

So, oxidation and reduction can also be defined in terms of electron loss and gain.

Oxidation is loss of electrons. **Reduction** is gain of electrons. **Redox** is electron transfer.

Remember **OIL RIG**! **O**xidation **I**s **L**oss, **R**eduction **I**s **G**ain.

ELECTROLYSIS

Electrolytic reactions are examples of redox reactions.

Anode reactions are **oxidations**.

At the positive anode, electrons are lost from anions:

e.g. $2Cl^- \rightarrow Cl_2 + 2e^-$
or $2O^{2-} \rightarrow O_2 + 4e^-$

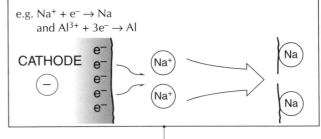

Cathode reactions are **reductions**.
At the negative cathode, electrons are gained by cations:

e.g. $Na^+ + e^- \rightarrow Na$
and $Al^{3+} + 3e^- \rightarrow Al$

CATHODE

DISPLACEMENT REACTIONS

Displacement reactions are also redox reactions. For example:

A more reactive metal is oxidized, while a less reactive one is reduced.

$Fe + Cu^{2+} \rightarrow Fe^{2+} + Cu$

A more reactive non-metal is reduced while a less reactive one is oxidized:

$Cl_2 + 2I^- \rightarrow 2Cl^- + I_2$

Reversible reactions

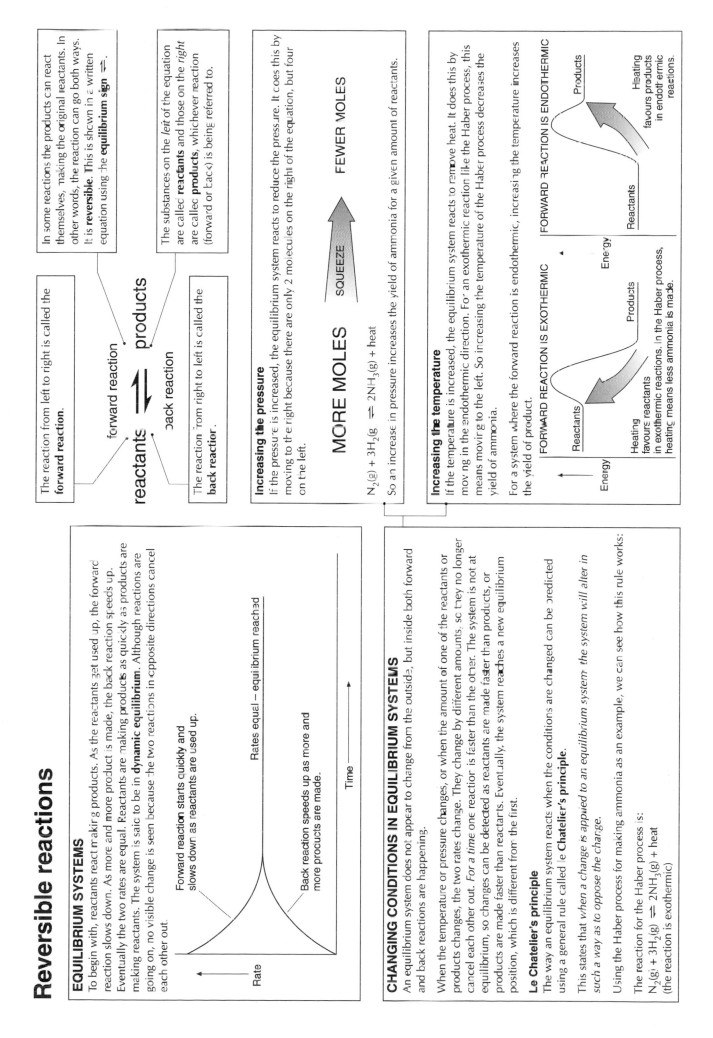

In some reactions the products can react themselves, making the original reactants. In other words, the reaction can go both ways. It is **reversible**. This is shown in a written equation using the **equilibrium sign** ⇌.

The substances on the *left* of the equation are called **reactants** and those on the *right* are called **products**, whichever reaction (forward or back) is being referred to.

The reaction from left to right is called the **forward reaction**.

The reaction from right to left is called the **back reaction**.

$$\text{reactants} \rightleftharpoons \text{products}$$

forward reaction

back reaction

EQUILIBRIUM SYSTEMS

To begin with, reactants react making products. As the reactants get used up, the forward reaction slows down. As more and more product is made, the back reaction speeds up. Eventually the two rates are equal. Reactants are making products as quickly as products are making reactants. The system is said to be in **dynamic equilibrium**. Although reactions are going on, no visible change is seen because the two reactions in opposite directions cancel each other out.

Rates equal – equilibrium reached

Forward reaction starts quickly and slows down as reactants are used up.

Back reaction speeds up as more and more products are made.

Rate

Time

Increasing the pressure

If the pressure is increased, the equilibrium system reacts to reduce the pressure. It does this by moving to the right because there are only 2 molecules on the right of the equation, but four on the left.

MORE MOLES   FEWER MOLES

SQUEEZE

$$N_2(g) + 3H_2(g) \rightleftharpoons 2NH_3(g) + heat$$

So an increase in pressure increases the yield of ammonia for a given amount of reactants.

Increasing the temperature

If the temperature is increased, the equilibrium system reacts to remove heat. It does this by moving in the endothermic direction. For an exothermic reaction like the Haber process, this means moving to the left. So increasing the temperature of the Haber process decreases the yield of ammonia.

For a system where the forward reaction is endothermic, increasing the temperature increases the yield of product.

FORWARD REACTION IS EXOTHERMIC

Energy

Reactants

Products

Heating favours reactants in exothermic reactions. In the Haber process, heating means less ammonia is made.

FORWARD REACTION IS ENDOTHERMIC

Energy

Products

Reactants

Heating favours products in endothermic reactions.

CHANGING CONDITIONS IN EQUILIBRIUM SYSTEMS

An equilibrium system does not appear to change from the outside, but inside both forward and back reactions are happening.

When the temperature or pressure changes, or when the amount of one of the reactants or products changes, the two rates change. They change by different amounts, so they no longer cancel each other out. *For a time* one reaction is faster than the other. The system is not at equilibrium, so changes can be detected as reactants are made faster than products, or products are made faster than reactants. Eventually, the system reaches a new equilibrium position, which is different from the first.

Le Chatelier's principle

The way an equilibrium system reacts when the conditions are changed can be predicted using a general rule called le **Chatelier's principle**.

This states that *when a change is applied to an equilibrium system the system will alter in such a way as to oppose the change*.

Using the Haber process for making ammonia as an example, we can see how this rule works:

The reaction for the Haber process is:
$$N_2(g) + 3H_2(g) \rightleftharpoons 2NH_3(g) + heat$$
(the reaction is exothermic)

Measuring rates of reaction

Rates of reaction

The rate of a chemical reaction is the **amount of reactant used up, or product made, in a given time**. Do not use the words 'speed' or 'how fast'.

The rate of a reaction can be measured in one of two general ways.

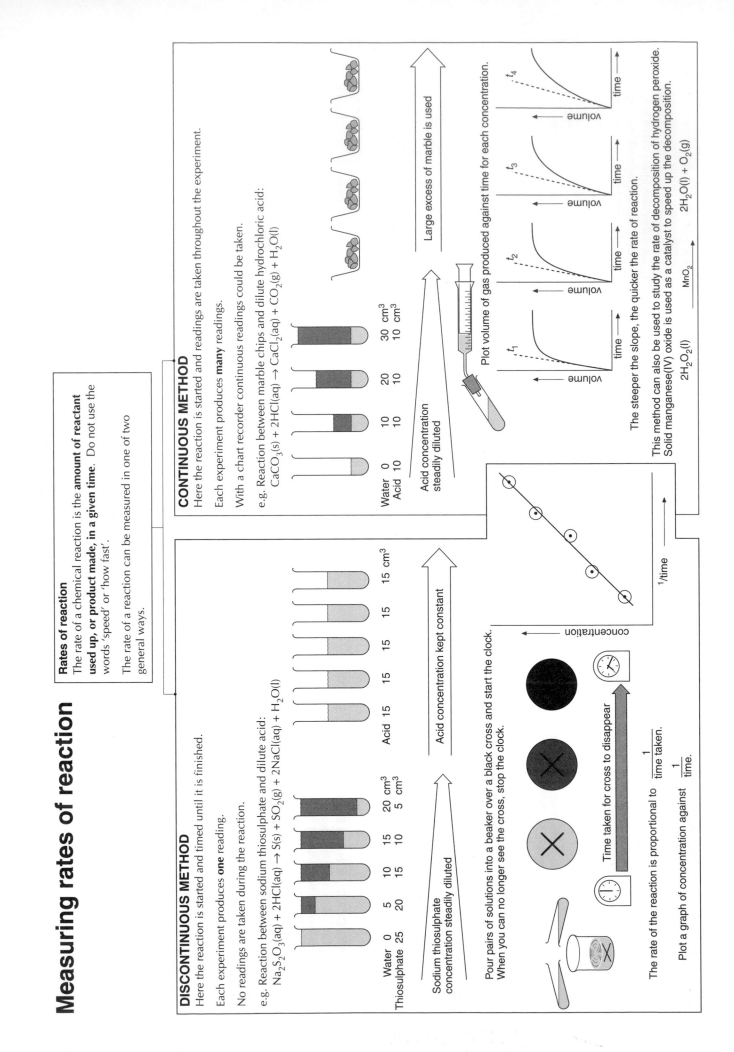

DISCONTINUOUS METHOD

Here the reaction is started and timed until it is finished.

Each experiment produces **one** reading.

No readings are taken during the reaction.

e.g. Reaction between sodium thiosulphate and dilute acid:
$Na_2S_2O_3(aq) + 2HCl(aq) \rightarrow S(s) + SO_2(g) + 2NaCl(aq) + H_2O(l)$

Water	0	5	10	15	20 cm³
Thiosulphate	25	20	15	10	5 cm³

Sodium thiosulphate concentration steadily diluted

Acid	15	15	15	15	15 cm³

Acid concentration kept constant

Pour pairs of solutions into a beaker over a black cross and start the clock. When you can no longer see the cross, stop the clock.

Time taken for cross to disappear

The rate of the reaction is proportional to $\dfrac{1}{\text{time taken}}$.

Plot a graph of concentration against $\dfrac{1}{\text{time}}$.

concentration

1/time

CONTINUOUS METHOD

Here the reaction is started and readings are taken throughout the experiment.

Each experiment produces **many** readings.

With a chart recorder continuous readings could be taken.

e.g. Reaction between marble chips and dilute hydrochloric acid:
$CaCO_3(s) + 2HCl(aq) \rightarrow CaCl_2(aq) + CO_2(g) + H_2O(l)$

Water	0	10	20	30 cm³
Acid	10	10	10	10 cm³

Acid concentration steadily diluted

Large excess of marble is used

Plot volume of gas produced against time for each concentration.

t_1 t_2 t_3 t_4

volume   time

The steeper the slope, the quicker the rate of reaction.

This method can also be used to study the rate of decomposition of hydrogen peroxide. Solid manganese(IV) oxide is used as a catalyst to speed up the decomposition.

$$2H_2O_2(l) \xrightarrow{MnO_2} 2H_2O(l) + O_2(g)$$

Factors affecting reaction rates

FACTORS AFFECTING RATE

The rate of a chemical reaction is affected by:

- **the concentration of the reactants**: increasing the concentration increases the rate
- **the pressure in gas state reactions**: increasing the pressure increases the rate
- **the surface area of solid reactions**: increasing the surface area increases the rate
- **the temperature of the reacting system**: increasing the temperature increases the rate enormously. For a typical chemical reaction, the rate *doubles for every 10°C rise in temperature*.
- **the addition of a catalyst**: adding a suitable catalyst increases the rate. A catalyst is a substance which *speeds up the rate of a reaction without itself being used up during the reaction*.

COLLISION THEORY

These facts can be explained using the **collision theory** which states:

For substances to react, their particles
- must collide
- with enough energy to break existing bonds.

Activation energy

Activation energy of catalysed reaction

Reactants

Products

Energy

Concentration

Increasing the concentration increases the number of particles.
Increasing the number of particles increases the number of collisions.
Increasing the number of collisions increases the number of successful collisions.
This increases the rate of reaction.

CONCENTRATION

Pressure

Increasing the pressure means that the gas molecules are squashed into a smaller volume.
The same amount of gas in a smaller volume has a greater concentration.
So the argument above applies. There will be more collisions, so there will be more successful collisions, so the rate will increase.

PRESSURE

CRUSH

Surface area and particle size

Only the particles on the surface of a solid are exposed to collisions.
Breaking up the solid makes new surfaces which are exposed to collisions.
So there are more collisions with the surface, which increases the reaction rate.

Temperature

Not all collisions between reactants succeed in making products. Only those collisions with enough energy to break bonds in the reactants will lead to a reaction. The energy a collision needs to be successful is called the **activation energy**. Increasing the temperature of the reaction means more particles have the activation energy. This means more collisions will be successful, so the rate of reaction increases.

Catalyst

A catalyst allows the reaction to go by a different pathway with a lower activation energy. More particles will have this lower activation energy, and so more collisions will be successful. More successful collisions means a higher rate. e.g. iron is added as catalyst in the Haber process for making ammonia; vanadium(V) oxide, V_2O_5, is added as a catalyst in the Contact process for making sulphuric acid; manganese(IV) oxide, MnO_2, catalyses the decomposition of hydrogen peroxide in the lab.

Reactions involving enzymes

Enzyme structure

Enzymes are covalent macromolecules made of amino acids bonded together making a protein. The amino acid chains are folded and twisted forming globular proteins with specific reactive sites.

Reactive site

Enzyme catalysis

Enzymes catalyse specific reactions in living cells by providing a reaction pathway with a lower activation energy.

Uncatalysed reaction

Catalysed reaction

Energy

pH dependence

The shape of an enzyme depends on the hydrogen bonds between different parts of the amino acid chain, and hydrogen bonds are affected by pH. So each enzyme works best at a particular pH.

Pepsin in the stomach works best in acid conditions.

Trypsin in the lower gut works best in alkaline conditions.

Rate

pH — 1 2 3 4 5 6 7 8 9 10 11 12 13

ENZYMES

- are all proteins found in living cells
- catalyse reactions in cells
- can each catalyse only one specific reaction
- act within a narrow pH range
- act within a narrow temperature range

Temperature dependence

At this temperature the protein molecule begins to lose its shape. It is denatured.

The reactants no longer fit the reactive site. Catalysis stops. This is irreversible. Cooling will not restore the catalytic activity of the enzyme.

Rate increases with temperature because there are more successful collisions.

Rate

Temperature (in °C) — 10 20 30 40 50 60

Enzyme lock and key mechanism

Reactant molecules fit into the reactive site of an enzyme like a key into a lock. Each reactive site only fits a particular reactant, so each enzyme catalyses only one reaction. Once in the site, the reactants react more quickly to form products.

Enzyme   Reactant

Products

Enzyme

USES OF ENZYMES

Digestion

Digestive enzymes convert insoluble substances in food into soluble substances which the body can absorb. They do this by breaking certain bonds one after another. Each bond is broken by a particular enzyme, and each enzyme needs particular conditions. This is why the pH changes through the gut.

Fermentation and baking

Fermentation is the anaerobic decomposition of simple carbohydrates into alcohol and carbon dioxide:

$$C_6H_{12}O_6 \rightarrow 2CO_2 + 2CH_3CH_2OH$$

This reaction is used in the brewing and wine making industries to make alcoholic drinks. In the baking industry the carbon dioxide produced makes the dough rise.

Yoghurt

Milk at 40°C is inoculated with bacteria. Enzymes in the bacteria convert the lactose in milk to lactic acid, which makes the milk thicken and curdle.

Detergents

Enzymes are added to detergents to help digest protein stains. Dirty washing is soaked in tepid water with a pH between 4 and 8, before being washed normally in hot water.

The enzymes used in detergents can sometimes cause allergic reactions such as asthma.

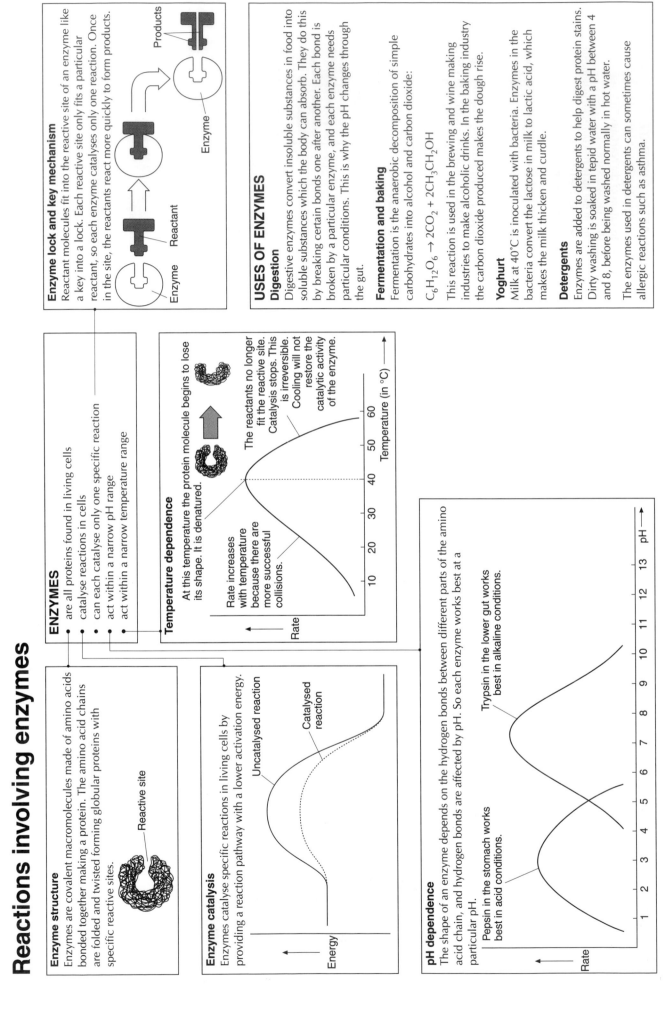

Formation of Earth's atmosphere

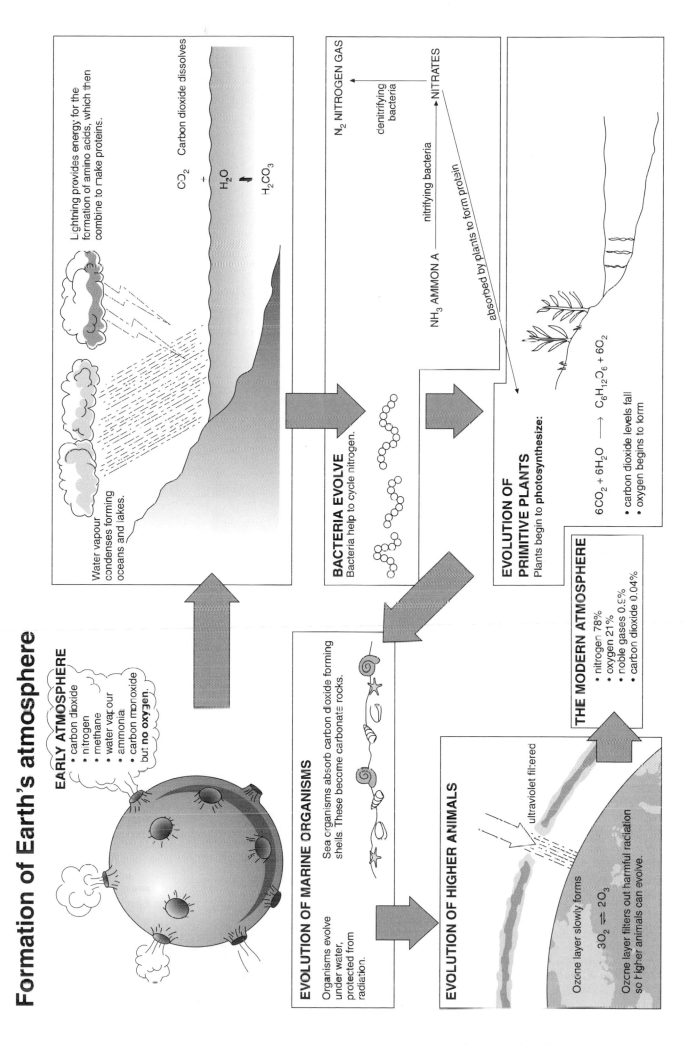

EARLY ATMOSPHERE
- carbon dioxide
- nitrogen
- methane
- water vapour
- ammonia
- carbon monoxide
but: **no oxygen.**

Lightning provides energy for the formation of amino acids, which then combine to make proteins.

Water vapour condenses forming oceans and lakes.

CO_2 Carbon dioxide dissolves

$CO_2 + H_2O \rightleftharpoons H_2CO_3$

N_2 NITROGEN GAS

denitrifying bacteria

NITRATES

NH_3 AMMONIA — nitrifying bacteria

absorbed by plants to form protein

BACTERIA EVOLVE
Bacteria help to cycle nitrogen.

EVOLUTION OF MARINE ORGANISMS

Organisms evolve under water, protected from radiation.

Sea organisms absorb carbon dioxide forming shells. These become carbonate rocks.

EVOLUTION OF PRIMITIVE PLANTS
Plants begin to photosynthesize:

$6CO_2 + 6H_2O \longrightarrow C_6H_{12}O_6 + 6O_2$
- carbon dioxide levels fall
- oxygen begins to form

EVOLUTION OF HIGHER ANIMALS

ultraviolet filtered

Ozone layer slowly forms

$3O_2 \rightleftharpoons 2O_3$

Ozone layer filters out harmful radiation so higher animals can evolve.

THE MODERN ATMOSPHERE
- nitrogen 78%
- oxygen 21%
- noble gases 0.9%
- carbon dioxide 0.04%

Changes in the atmosphere

CARBON DIOXIDE AND THE GREENHOUSE EFFECT

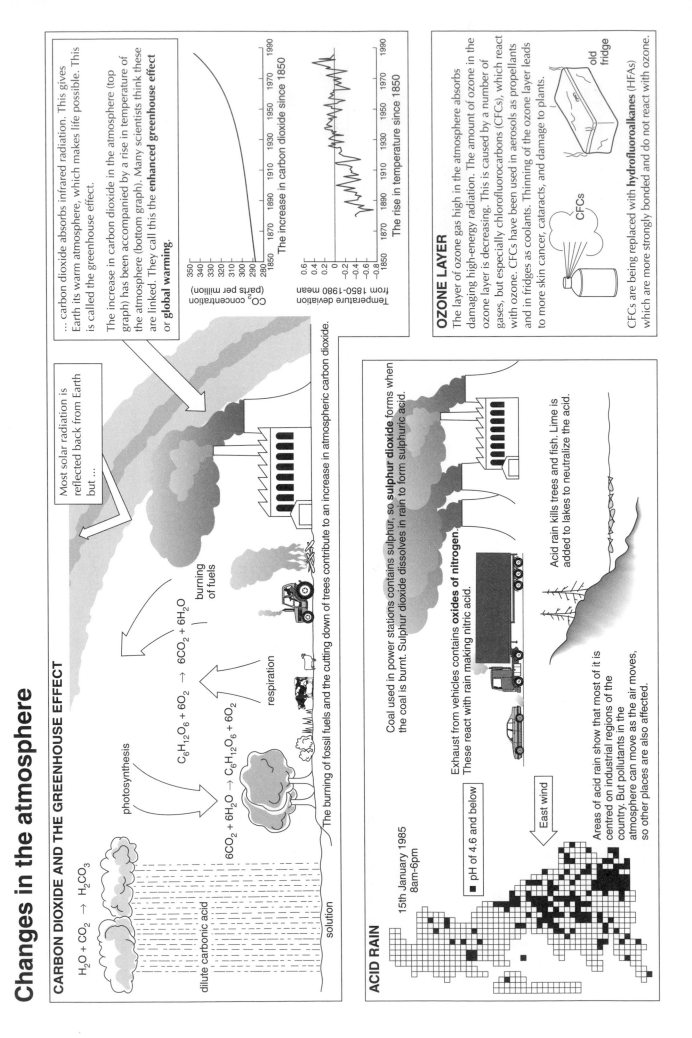

... carbon dioxide absorbs infrared radiation. This gives Earth its warm atmosphere, which makes life possible. This is called the greenhouse effect.

The increase in carbon dioxide in the atmosphere (top graph) has been accompanied by a rise in temperature of the atmosphere (bottom graph). Many scientists think these are linked. They call this the **enhanced greenhouse effect** or **global warming**.

Most solar radiation is reflected back from Earth but ...

$H_2O + CO_2 \rightarrow H_2CO_3$

dilute carbonic acid

solution

$6CO_2 + 6H_2O \rightarrow C_6H_{12}O_6 + 6O_2$

photosynthesis

$C_6H_{12}O_6 + 6O_2 \rightarrow 6CO_2 + 6H_2O$

respiration

burning of fuels

The burning of fossil fuels and the cutting down of trees contribute to an increase in atmospheric carbon dioxide.

CO$_2$ concentration (parts per million)

The increase in carbon dioxide since 1850

Temperature deviation from 1850-1980 mean

The rise in temperature since 1850

OZONE LAYER

The layer of ozone gas high in the atmosphere absorbs damaging high-energy radiation. The amount of ozone in the ozone layer is decreasing. This is caused by a number of gases, but especially chlorofluorocarbons (CFCs), which react with ozone. CFCs have been used in aerosols as propellants and in fridges as coolants. Thinning of the ozone layer leads to more skin cancer, cataracts, and damage to plants.

old fridge

CFCs

CFCs are being replaced with **hydrofluoroalkanes** (HFAs) which are more strongly bonded and do not react with ozone.

ACID RAIN

Coal used in power stations contains sulphur, so **sulphur dioxide** forms when the coal is burnt. Sulphur dioxide dissolves in rain to form sulphuric acid.

Exhaust from vehicles contains **oxides of nitrogen**. These react with rain making nitric acid.

Acid rain kills trees and fish. Lime is added to lakes to neutralize the acid.

15th January 1985 8am-6pm

■ pH of 4.6 and below

East wind

Areas of acid rain show that most of it is centred on industrial regions of the country. But pollutants in the atmosphere can move as the air moves, so other places are also affected.

Products from air

FRACTIONAL DISTILLATION OF AIR

Air is a mixture so is separated by a physical process.

Because b.p.s are close together, **fractional distillation** is used.

Nitrogen b.p. −196°C
Used for:
- shrink fitting
- keeping flammable materials safe
- food freezing
- food packaging

Argon b.p. −186°C
Used for:
- arc welding
- light bulbs

Oxygen b.p. −183°C
Used for:
- steel making
- aerating rivers
- oxyacetylene cutting
- medical uses

Dried, filtered air

FIXATION OF NITROGEN AND THE HABER PROCESS

- Plants are made of cells – cells have nuclei – nuclei contain proteins – proteins are made from amino acids – amino acids contain nitrogen.
- **Fixation** is the conversion of unreactive atmospheric nitrogen into reactive compounds that plants can absorb.

NATURAL FIXATION

- thunderstorms

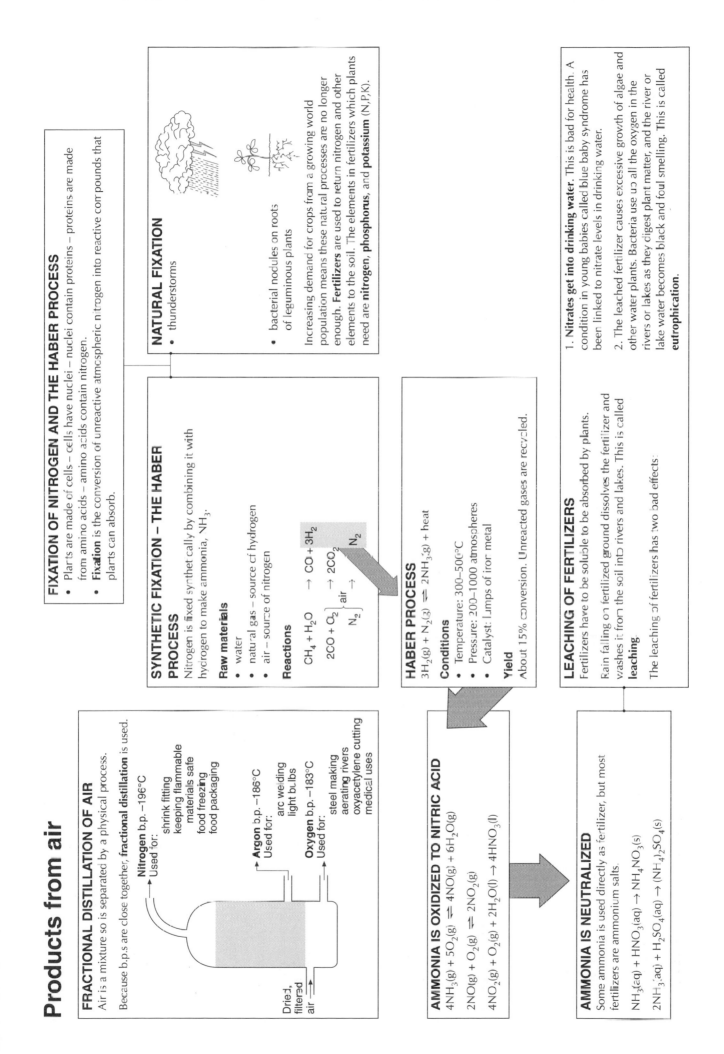

- bacterial nodules on roots of leguminous plants

Increasing demand for crops from a growing world population means these natural processes are no longer enough. **Fertilizers** are used to return nitrogen and other elements to the soil. The elements in fertilizers which plants need are **nitrogen**, **phosphorus**, and **potassium** (N,P,K).

SYNTHETIC FIXATION – THE HABER PROCESS

Nitrogen is fixed synthetically by combining it with hydrogen to make ammonia, NH_3.

Raw materials
- water
- natural gas – source of hydrogen
- air – source of nitrogen

Reactions

$CH_4 + H_2O \rightarrow CO + 3H_2$

$2CO + O_2 \rightarrow 2CO_2$ } air

$N_2 \rightarrow$

N_2

HABER PROCESS

$3H_2(g) + N_2(g) \rightleftharpoons 2NH_3(g) + heat$

Conditions
- Temperature: 300–500°C
- Pressure: 200–1000 atmospheres
- Catalyst: lumps of iron metal

Yield
About 15% conversion. Unreacted gases are recycled.

AMMONIA IS OXIDIZED TO NITRIC ACID

$4NH_3(g) + 5O_2(g) \rightleftharpoons 4NO(g) + 6H_2O(g)$

$2NO(g) + O_2(g) \rightleftharpoons 2NO_2(g)$

$4NO_2(g) + O_2(g) + 2H_2O(l) \rightarrow 4HNO_3(l)$

AMMONIA IS NEUTRALIZED

Some ammonia is used directly as fertilizer, but most fertilizers are ammonium salts.

$NH_3(aq) + HNO_3(aq) \rightarrow NH_4NO_3(s)$

$2NH_3(aq) + H_2SO_4(aq) \rightarrow (NH_4)_2SO_4(s)$

LEACHING OF FERTILIZERS

Fertilizers have to be soluble to be absorbed by plants.

Rain falling on fertilized ground dissolves the fertilizer and washes it from the soil into rivers and lakes. This is called **leaching**

The leaching of fertilizers has two bad effects:

1. **Nitrates get into drinking water**. This is bad for health. A condition in young babies called blue baby syndrome has been linked to nitrate levels in drinking water.

2. The leached fertilizer causes excessive growth of algae and other water plants. Bacteria use up all the oxygen in the rivers or lakes as they digest plant matter, and the river or lake water becomes black and foul smelling. This is called **eutrophication.**

Choosing conditions for the Haber process

REACTION

$$N_2(g) + 3H_2(g) \rightleftharpoons 2 NH_3(g) + heat$$

$:N \equiv N:$ $H—H$ $\begin{array}{c} H \\ :N—H \\ H \end{array}$

very strongly bonded
944 kJ mol⁻¹

quite a strong bond
436 kJ mol⁻¹

weaker bonds than in the reactants
388 kJ mol⁻¹

THE PROBLEM

The reactant molecules (N_2 and H_2) are very strongly bonded. Only collisions with a huge amount of energy will be successful, because the activation energy is very high. This means that the rate of the forward reaction making ammonia will be very slow unless the temperature is high. But if the temperature is high, the relatively weak bonds in ammonia will break and the back reaction will speed up. So a high temperature is needed to supply the activation energy, but a high temperature makes the product decompose more quickly.

LE CHATELIER

Le Chatelier's principle predicts that the best yield of ammonia will be produced under high pressure and low temperature.

Pressure

There are more moles of reactants than products, so high pressure favours products.

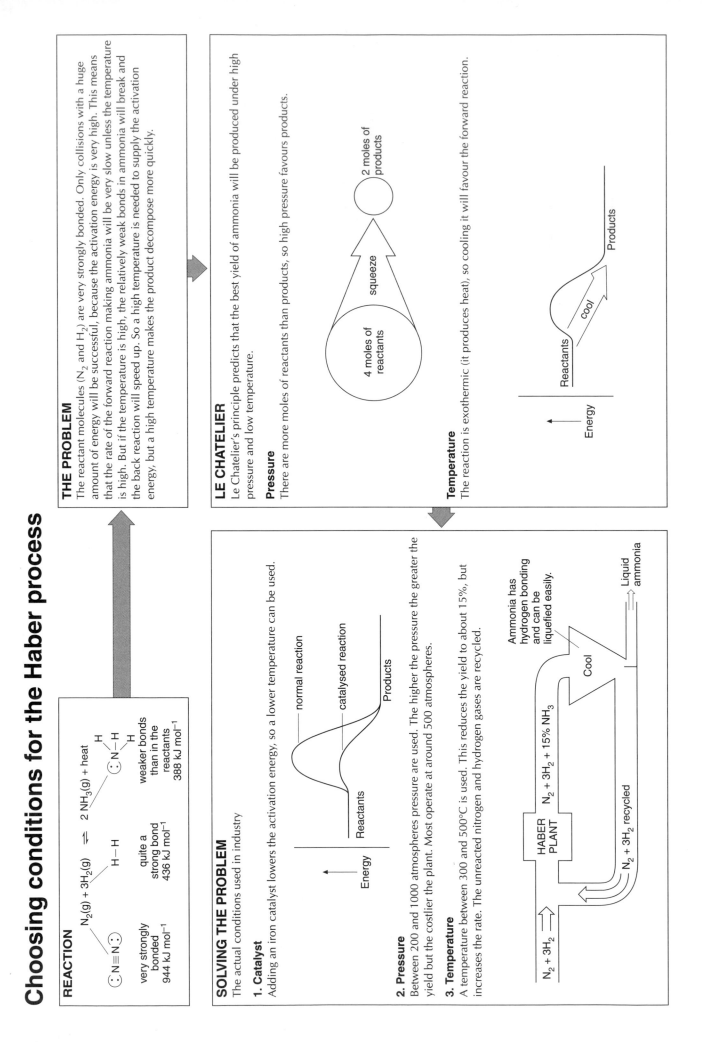

4 moles of reactants

squeeze

2 moles of products

Temperature

The reaction is exothermic (it produces heat), so cooling it will favour the forward reaction.

Energy

Reactants

cool

Products

SOLVING THE PROBLEM

The actual conditions used in industry

1. Catalyst

Adding an iron catalyst lowers the activation energy, so a lower temperature can be used.

Energy

Reactants

normal reaction

catalysed reaction

Products

2. Pressure

Between 200 and 1000 atmospheres pressure are used. The higher the pressure the greater the yield but the costlier the plant. Most operate at around 500 atmospheres.

3. Temperature

A temperature between 300 and 500°C is used. This reduces the yield to about 15%, but increases the rate. The unreacted nitrogen and hydrogen gases are recycled.

$N_2 + 3H_2$

HABER PLANT

$N_2 + 3H_2 + 15\% NH_3$

Cool

Ammonia has hydrogen bonding and can be liquefied easily.

Liquid ammonia

$N_2 + 3H_2$ recycled

Natural cycles

Natural cycles are very important because:
1. They maintain various elements so that they can be used again and again.
2. They maintain a balance or equilibrium in different parts of the cycle by replacing an element as it is used up.

When these cycles are disturbed by human activities, environmental problems build up. Either a resource runs out, or some substance builds up in concentration and causes pollution.

NITROGEN CYCLE

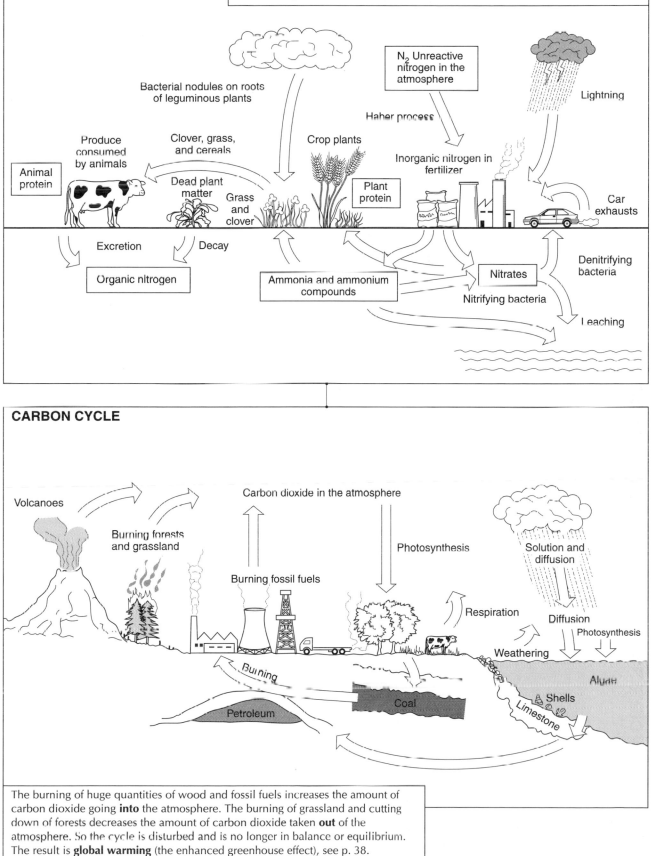

CARBON CYCLE

The burning of huge quantities of wood and fossil fuels increases the amount of carbon dioxide going **into** the atmosphere. The burning of grassland and cutting down of forests decreases the amount of carbon dioxide taken **out** of the atmosphere. So the cycle is disturbed and is no longer in balance or equilibrium. The result is **global warming** (the enhanced greenhouse effect), see p. 38.

The rock cycle and different kinds of rocks

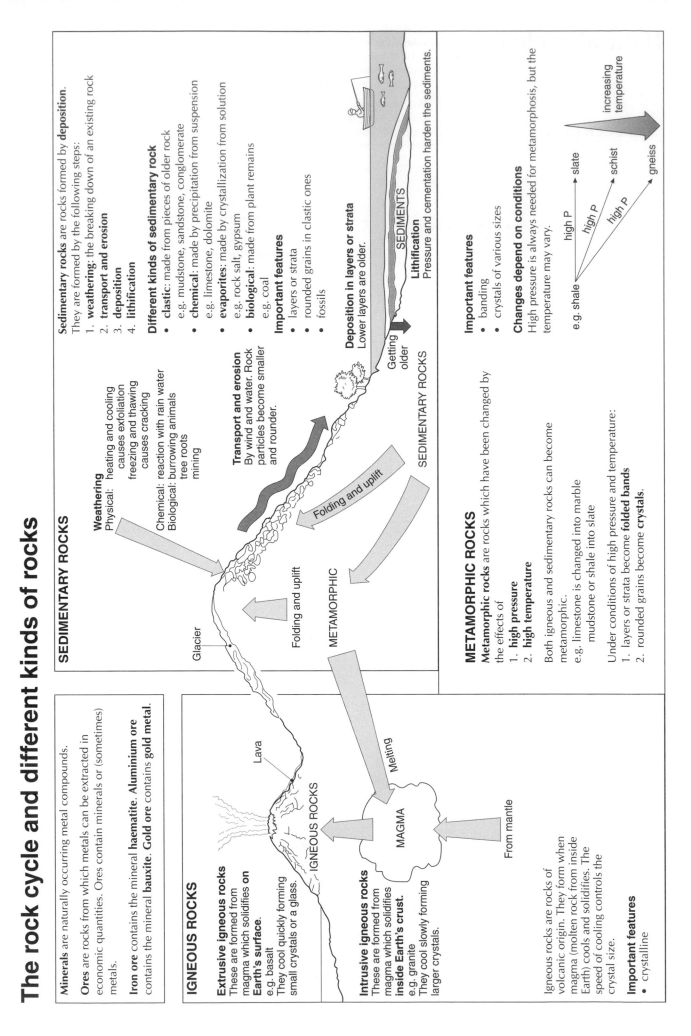

SEDIMENTARY ROCKS

Sedimentary rocks are rocks formed by **deposition**.
They are formed by the following steps:
1. **weathering**: the breaking down of an existing rock
2. **transport and erosion**
3. **deposition**
4. **lithification**

Different kinds of sedimentary rock
- **clastic**: made from pieces of older rock
  e.g. mudstone, sandstone, conglomerate
- **chemical**: made by precipitation from suspension
  e.g. limestone, dolomite
- **evaporites**: made by crystallization from solution
  e.g. rock salt, gypsum
- **biological**: made from plant remains
  e.g. coal

Important features
- layers or strata
- rounded grains in clastic ones
- fossils

Deposition in layers or strata
Lower layers are older.

SEDIMENTS

Lithification
Pressure and cementation harden the sediments.

Weathering
Physical: heating and cooling causes exfoliation
freezing and thawing causes cracking

Chemical: reaction with rain water
Biological: burrowing animals
tree roots
mining

Transport and erosion
By wind and water. Rock particles become smaller and rounder.

Getting older

SEDIMENTARY ROCKS

Folding and uplift

METAMORPHIC

Glacier

IGNEOUS ROCKS

Minerals are naturally occurring metal compounds.

Ores are rocks from which metals can be extracted in economic quantities. Ores contain minerals or (sometimes) metals.

Iron ore contains the mineral **haematite**. **Aluminium ore** contains the mineral **bauxite**. **Gold ore** contains **gold metal**.

IGNEOUS ROCKS

Extrusive igneous rocks
These are formed from magma which solidifies **on Earth's surface.**
e.g. basalt
They cool quickly forming small crystals or a glass.

Intrusive igneous rocks
These are formed from magma which solidifies **inside Earth's crust.**
e.g. granite
They cool slowly forming larger crystals.

Igneous rocks are rocks of volcanic origin. They form when magma (molten rock from inside Earth) cools and solidifies. The speed of cooling controls the crystal size.

Important features
- crystalline

Lava

IGNEOUS ROCKS

MAGMA

Melting

From mantle

METAMORPHIC ROCKS

Metamorphic rocks are rocks which have been changed by the effects of
1. **high pressure**
2. **high temperature**

Both igneous and sedimentary rocks can become metamorphic.
e.g. limestone is changed into marble
mudstone or shale into slate

Under conditions of high pressure and temperature:
1. layers or strata become **folded bands**
2. rounded grains become **crystals.**

Important features
- banding
- crystals of various sizes

Changes depend on conditions
High pressure is always needed for metamorphism, but the temperature may vary.

e.g. shale → high P → slate → high P → schist → high P → gneiss

increasing temperature

The structure of Earth and the theory of tectonic plates

TECTONIC PLATE THEORY

This states that the continents are made of less dense crustal rock floating on plates of the mantle. These plates move because there are convection currents in the inner part of Earth.

Evidence for tectonic plate theory

1. Continental coastlines fit together like bits of a jigsaw.
   e.g. the coasts of Africa and South America match each other.

2. There are similar rocks on either side of the oceans which match when the continents are moved together.
   e.g. similar rocks and fossils are found in Africa and South America.

3. Geological evidence of climatic change suggests that plates have moved.
   e.g. desert sands, tropical fossils, and glacial deposits show that Britain has moved on Earth's surface.

4. Position of mountain chains and island arcs.
   e.g. Japan and the Philippines show where plate margins are meeting.

5. Magnetic record preserved in rocks: magnetic stripes and wandering north pole.
   e.g. parallel lines of magnetism on either side of mid-ocean ridges show that there has been sea-floor spreading.

EVIDENCE FOR EARTH'S STRUCTURE

1. **Volcanoes**
   They indicate that the inside of Earth is hot and under pressure.

2. **Earthquakes**
   A pattern of shock waves spreads out through Earth from an earthquake. The pattern made by the waves tells us about the structure of Earth.

EARTH'S STRUCTURE

Studying earthquake waves tells us that Earth has a layered structure made of:
- a **very thin crust** a few tens of kilometres thick
- the **mantle** of underlying rock about 5000 km thick
- a **liquid core** about 2000 km thick
- a **solid core** about 1000 km thick in the centre.

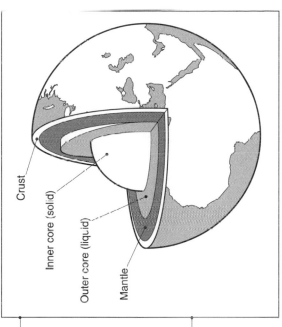

Crust

Inner core (solid)

Outer core (liquid)

Mantle

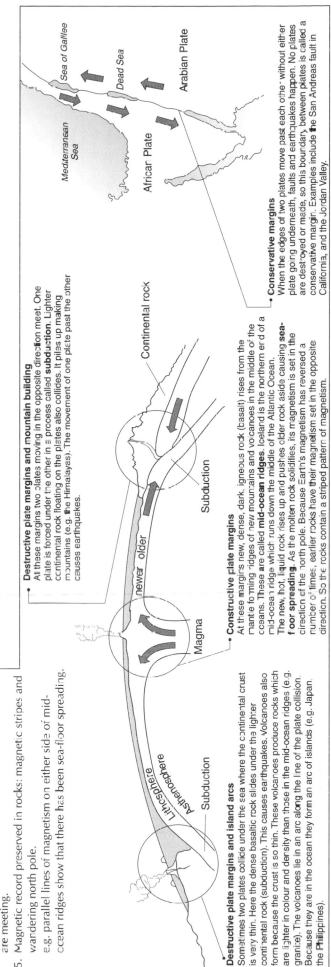

Sea of Galilee

Dead Sea

Arabian Plate

Mediterranean Sea

African Plate

- **Conservative margins**
  When the edges of two plates move past each other without either plate going underneath, faults and earthquakes happen. No plates are destroyed or made, so this boundary between plates is called a conservative margin. Examples include the San Andreas fault in California, and the Jordan Valley.

- **Destructive plate margins and mountain building**
  At these margins two plates moving in the opposite direction meet. One plate is forced under the other in a process called **subduction**. Lighter continental rock floating on the plates also collides. It piles up making mountains (e.g. the Himalayas). The movement of one plate past the other causes earthquakes.

Continental rock

Subduction

newer   older

Magma

Lithosphere

Asthenosphere

Subduction

- **Constructive plate margins**
  At these margins new, dense, dark, igneous rock (basalt) rises from the mantle forming ridges of new mountains and volcanoes in the middle of the oceans. These are called **mid-ocean ridges**. Iceland is the northern end of a mid-ocean ridge which runs down the middle of the Atlantic Ocean. The new, hot, liquid rock rises up and pushes older rock aside causing **sea-floor spreading**. As the molten rock solidifies, its magnetism is set in the direction of the north pole. Because Earth's magnetism has reversed a number of times, earlier rocks have their magnetism set in the opposite direction. So the rocks contain a striped pattern of magnetism.

Destructive plate margins and island arcs
Sometimes two plates collide under the sea where the continental crust is very thin. Here the dense basaltic rock slides under the lighter continental rock (subduction). This causes earthquakes. Volcanoes also form because the crust is so thin. These volcanoes produce rocks which are lighter in colour and density than those in the mid-ocean ridges (e.g. granite). The volcanoes lie in an arc along the line of the plate collision. Because they are in the ocean they form an arc of islands (e.g. Japan and the Philippines).

Materials from rocks: limestone and its uses

BUILDING MATERIAL

Limestone is a **hard sedimentary rock**. It is used for road foundations and to make buildings.

MAKING CEMENT

Limestone and clay are heated together in a furnace to make cement.

$$4CaCO_3(s) + Al_2Si_2O_7(s) \rightarrow 2CaSiO_3(s) + Ca_2Al_2O_5(s) + 4CO_2(g)$$

cement

MAKING GLASS

Limestone, sand, and sodium carbonate are heated together to make glass. This **soda glass** is used in windows. Soda glasses have the follow range of compositions:

SiO_2	70–75%
Na_2O	12–16%
CaO	5–11%

If boron oxide is added to the reaction mixture, a **borosilicate glass** is produced. Borosilicate glasses are harder than soda glasses and can be heated without softening. Pyrex is a borosilicate glass.

Calcium carbonate rocks

LIMESTONE QUARRY

MINING AND QUARRYING

These processes:
* create ugly slag heaps and scar the landscape
* generate a lot of heavy traffic
* cause dust and smoke pollution

but also
* provide jobs and income
* provide valuable raw materials.

NEUTRALIZING LAKES

Limestone (calcium carbonate) is a **base**. It neutralizes lakes made acid by acid rain. Powdered limestone can be sprayed onto lakes for this purpose.

MAKING LIME FOR AGRICULTURE

Metal carbonates decompose on heating. Heating limestone produces **quicklime** which reacts with water to make **slaked lime**.

$$CaCO_3(s) \xrightarrow{\text{heat}} CaO(s) + CO_2(g)$$
quicklime

$$CaO(s) + H_2O(l) \rightarrow Ca(OH)_2(s)$$
slaked lime

MAKING IRON

Limestone is mixed with iron ore and coke to form the 'charge' which is loaded into a **blast furnace** (see p. 45). The limestone reacts with the high melting point non-metal impurities in the iron ore. It forms a molten slag which floats on top of the iron.

Iron and steel

MANUFACTURE OF IRON

There is less iron than aluminium in Earth's crust, but it is easier and cheaper to extract.

The main ore is **haematite**, Fe_2O_3.

Iron is below zinc in the reactivity list so it forms fairly reactive compounds. Iron is extracted from haematite by **reducing it with carbon in a blast furnace**.

The **charge** is loaded into the top of the furnace.

The charge contains:
* iron ore – source of iron
* coke – fuel and reducing agent
* limestone – to form a **slag** by dissolving the high melting point non-metal impurities.

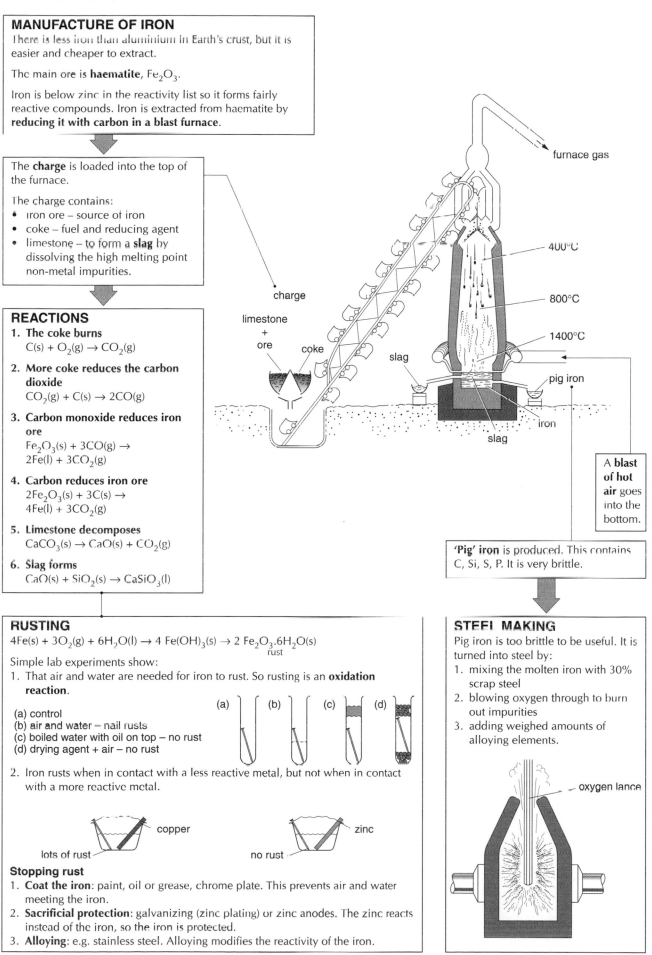

furnace gas

charge

limestone + ore

coke

slag

400°C

800°C

1400°C

pig iron

iron

slag

A **blast of hot air** goes into the bottom.

REACTIONS

1. **The coke burns**
   $C(s) + O_2(g) \rightarrow CO_2(g)$

2. **More coke reduces the carbon dioxide**
   $CO_2(g) + C(s) \rightarrow 2CO(g)$

3. **Carbon monoxide reduces iron ore**
   $Fe_2O_3(s) + 3CO(g) \rightarrow 2Fe(l) + 3CO_2(g)$

4. **Carbon reduces iron ore**
   $2Fe_2O_3(s) + 3C(s) \rightarrow 4Fe(l) + 3CO_2(g)$

5. **Limestone decomposes**
   $CaCO_3(s) \rightarrow CaO(s) + CO_2(g)$

6. **Slag forms**
   $CaO(s) + SiO_2(s) \rightarrow CaSiO_3(l)$

'**Pig**' **iron** is produced. This contains C, Si, S, P. It is very brittle.

RUSTING

$4Fe(s) + 3O_2(g) + 6H_2O(l) \rightarrow 4 Fe(OH)_3(s) \rightarrow 2 Fe_2O_3.6H_2O(s)$
$\underbrace{}_{rust}$

Simple lab experiments show:
1. That air and water are needed for iron to rust. So rusting is an **oxidation reaction**.

   (a) control
   (b) air and water – nail rusts
   (c) boiled water with oil on top – no rust
   (d) drying agent + air – no rust

   (a)   (b)   (c)   (d)

2. Iron rusts when in contact with a less reactive metal, but not when in contact with a more reactive metal.

   copper — lots of rust

   zinc — no rust

Stopping rust
1. **Coat the iron**: paint, oil or grease, chrome plate. This prevents air and water meeting the iron.
2. **Sacrificial protection**: galvanizing (zinc plating) or zinc anodes. The zinc reacts instead of the iron, so the iron is protected.
3. **Alloying**: e.g. stainless steel. Alloying modifies the reactivity of the iron.

STEEL MAKING

Pig iron is too brittle to be useful. It is turned into steel by:
1. mixing the molten iron with 30% scrap steel
2. blowing oxygen through to burn out impurities
3. adding weighed amounts of alloying elements.

oxygen lance

Aluminium extraction

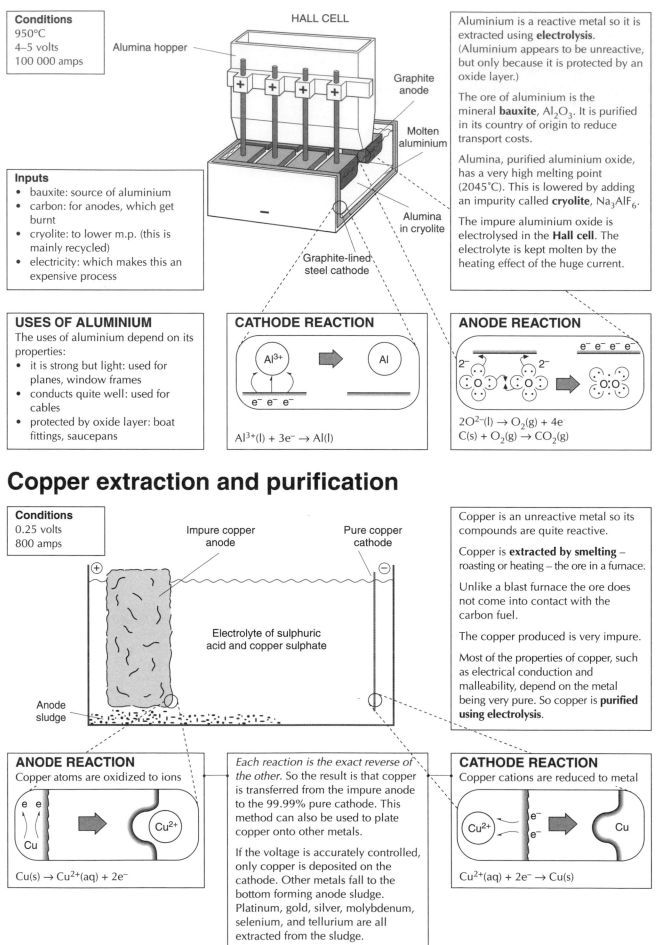

Conditions
950°C
4–5 volts
100 000 amps

HALL CELL

Alumina hopper

Graphite anode

Molten aluminium

Alumina in cryolite

Graphite-lined steel cathode

Inputs
- bauxite: source of aluminium
- carbon: for anodes, which get burnt
- cryolite: to lower m.p. (this is mainly recycled)
- electricity: which makes this an expensive process

Aluminium is a reactive metal so it is extracted using **electrolysis**. (Aluminium appears to be unreactive, but only because it is protected by an oxide layer.)

The ore of aluminium is the mineral **bauxite**, Al_2O_3. It is purified in its country of origin to reduce transport costs.

Alumina, purified aluminium oxide, has a very high melting point (2045°C). This is lowered by adding an impurity called **cryolite**, Na_3AlF_6.

The impure aluminium oxide is electrolysed in the **Hall cell**. The electrolyte is kept molten by the heating effect of the huge current.

USES OF ALUMINIUM
The uses of aluminium depend on its properties:
- it is strong but light: used for planes, window frames
- conducts quite well: used for cables
- protected by oxide layer: boat fittings, saucepans

CATHODE REACTION

$$Al^{3+}(l) + 3e^- \rightarrow Al(l)$$

ANODE REACTION

e^- e^- e^- e^-

$$2O^{2-}(l) \rightarrow O_2(g) + 4e^-$$
$$C(s) + O_2(g) \rightarrow CO_2(g)$$

Copper extraction and purification

Conditions
0.25 volts
800 amps

Impure copper anode

Pure copper cathode

Electrolyte of sulphuric acid and copper sulphate

Anode sludge

Copper is an unreactive metal so its compounds are quite reactive.

Copper is **extracted by smelting** – roasting or heating – the ore in a furnace:

Unlike a blast furnace the ore does not come into contact with the carbon fuel.

The copper produced is very impure.

Most of the properties of copper, such as electrical conduction and malleability, depend on the metal being very pure. So copper is **purified using electrolysis**.

ANODE REACTION
Copper atoms are oxidized to ions

$$Cu(s) \rightarrow Cu^{2+}(aq) + 2e^-$$

Each reaction is the exact reverse of the other. So the result is that copper is transferred from the impure anode to the 99.99% pure cathode. This method can also be used to plate copper onto other metals.

If the voltage is accurately controlled, only copper is deposited on the cathode. Other metals fall to the bottom forming anode sludge. Platinum, gold, silver, molybdenum, selenium, and tellurium are all extracted from the sludge.

CATHODE REACTION
Copper cations are reduced to metal

$$Cu^{2+}(aq) + 2e^- \rightarrow Cu(s)$$

Transition elements

TITANIUM

Extraction

Titanium is very expensive to extract because it is extracted by a batch process instead of a continuous one.

Titanium is displaced from purified titanium chloride by magnesium, a more reactive metal.

$$TiCl_4(l) + 2Mg(l) \rightarrow Ti(s) + 2 MgCl_2(l)$$

Uses

Titanium is light but very strong. It is used in making turbine blades, aeroplane parts, and artificial limbs.

Once a **continuous process** is started up, metal is made continuously until the process is stopped. In a **batch process**, the furnace is loaded and heated up, reaction takes place, then the furnace is cooled down and unloaded. This whole process is then repeated. This wastes energy and time.

The transition elements are found between Groups 2 and 3 of the periodic table. They have characteristic properties (see p. 18). They also have a wide range of uses which depend on these properties.

(see p. 18)

COPPER, SILVER, AND GOLD

These are all easy to extract, but silver and gold are rare and so expensive. They all show typical metallic properties (good electrical conductivity, malleable, ductile). Their uses follow from their properties.

Copper

- is widely used as an electrical conductor (cheap, good electrical conductivity). But it is now being replaced by aluminium – which is cheaper and lighter– for cables in the electricity grid.
- is used for pipes (unreactive, ductile)
- is used for coins (unreactive, has a distinctive colour)

Silver and gold

- are sometimes used to coat electrical contacts, where the heat of sparking would cause copper to oxidize (unreactive, good electrical conductivity)
- are used for jewellery (rare, precious, attractive colours)

ZINC, CHROMIUM, NICKEL, AND COPPER

The biggest disadvantage of iron and steel is that they rust. Replacing or repairing rusting structures costs huge amounts every year. Rusting can be prevented by:

- galvanizing or coating the iron with zinc
- alloying the iron with a less reactive metal (e.g. chromium or nickel)
- chrome-plating steel, which provides a decorative, unreactive surface. But the steel will rust badly if the chrome layer is scratched through, because chromium is less reactive than iron.

Brass is sometimes used instead of iron. It is an alloy of copper and zinc.

						Zn
Ti	Cr	Mn	Fe	Ni	Cu	
					Ag	
					Au	

IRON

Cast iron is brittle but hard because of the impurities it contains. It is used for making manhole covers.

Steel

When iron is purified and alloyed with small amounts of carbon, steel is made. The amount of carbon changes the properties of the steel.

- **mild steel** (0.09–0.2% carbon): malleable and not very hard; used for car bodies, cans, nails, wire
- **high carbon steel** (0.4–0.9% carbon): harder and less malleable; used for tools, masonry nails

---

Alloy steels

The properties of steel can be further changed by adding other transition metals:

- **stainless steel** is 18% chromium, 8% nickel
- **manganese steel** is very strong and hard; used for drill bits, cutting tools, springs.

Chemicals from salt

Common salt, sodium chloride, is found naturally:

1. dissolved in **sea water** (about 2.6% in a typical sample) from which it is extracted by **evaporation**;
2. as **rock salt**, a sedimentary evaporite. There are huge deposits of rock salt in Cheshire. These are extracted by **solution mining**.

Sodium is a very reactive metal so it is extracted by **electrolysis**. Chlorine is also very reactive. The cells used for electrolysis are designed to keep the very reactive products apart.

Uses of sodium chloride

- added to food to preserve it (ham)
- put on icy roads to melt the ice
- a source of sodium, chlorine, and sodium hydroxide

ELECTROLYSIS OF MOLTEN SODIUM CHLORIDE

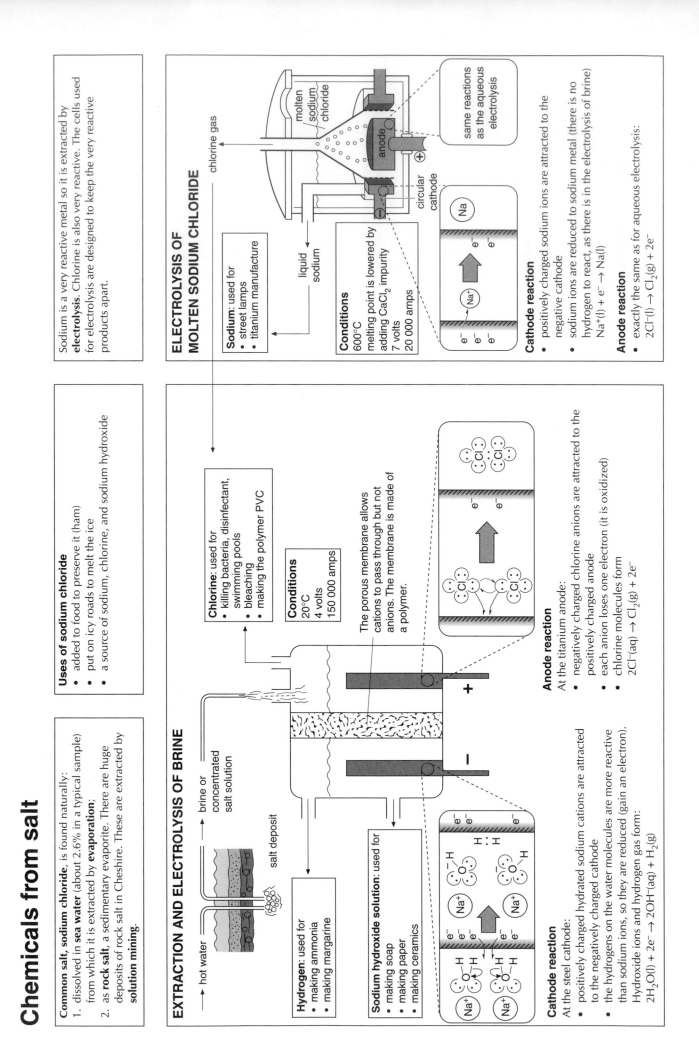

Sodium: used for
- street lamps
- titanium manufacture

Conditions
600°C
melting point is lowered by adding CaCl$_2$ impurity
7 volts
20 000 amps

Cathode reaction
- positively charged sodium ions are attracted to the negative cathode
- sodium ions are reduced to sodium metal (there is no hydrogen to react, as there is in the electrolysis of brine)
$$Na^+(l) + e^- \rightarrow Na(l)$$

Anode reaction
- exactly the same as for aqueous electrolysis:
$$2Cl^-(l) \rightarrow Cl_2(g) + 2e^-$$

same reactions as the aqueous electrolysis

EXTRACTION AND ELECTROLYSIS OF BRINE

Chlorine: used for
- killing bacteria, disinfectant, swimming pools
- bleaching
- making the polymer PVC

Conditions
20°C
4 volts
150 000 amps

The porous membrane allows cations to pass through but not anions. The membrane is made of a polymer.

Hydrogen: used for
- making ammonia
- making margarine

Sodium hydroxide solution: used for
- making soap
- making paper
- making ceramics

Cathode reaction
At the steel cathode:
- positively charged hydrated sodium cations are attracted to the negatively charged cathode
- the hydrogens on the water molecules are more reactive than sodium ions, so they are reduced (gain an electron).

Hydroxide ions and hydrogen gas form:
$$2H_2O(l) + 2e^- \rightarrow 2OH^-(aq) + H_2(g)$$

Anode reaction
At the titanium anode:
- negatively charged chlorine anions are attracted to the positively charged anode
- each anion loses one electron (it is oxidized)
- chlorine molecules form
$$2Cl^-(aq) \rightarrow Cl_2(g) + 2e^-$$

Chemicals from crude oil

EXTRACTION

Crude oil is:
- a mixture
- of saturated hydrocarbons
- whose boiling points are close together.

Crude oil or petroleum
- **a fossil fuel** – the remains of marine organisms
- **a non-renewable resource** – but new reserves continue to be found.

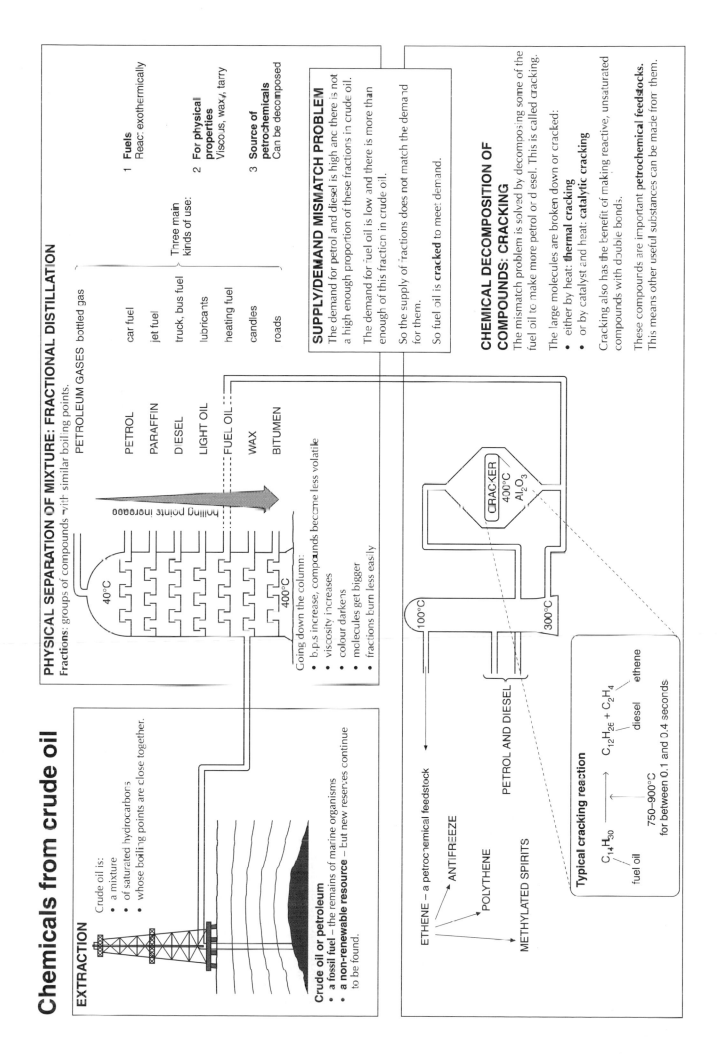

PHYSICAL SEPARATION OF MIXTURE: FRACTIONAL DISTILLATION

Fractions: groups of compounds with similar boiling points.

PETROLEUM GASES bottled gas

PETROL — car fuel

PARAFFIN — jet fuel

DIESEL — truck, bus fuel

LIGHT OIL — lubricants

FUEL OIL — heating fuel

WAX — candles

BITUMEN — roads

Three main kinds of use:

1 **Fuels**
React exothermically

2 **For physical properties**
Viscous, waxy, tarry

3 **Source of petrochemicals**
Can be decomposed

boiling points increase

40°C

400°C

Going down the column:
- b.p.s increase, compounds become less volatile
- viscosity increases
- colour darkens
- molecules get bigger
- fractions burn less easily

SUPPLY/DEMAND MISMATCH PROBLEM

The demand for petrol and diesel is high and there is not a high enough proportion of these fractions in crude oil.

The demand for fuel oil is low and there is more than enough of this fraction in crude oil.

So the supply of fractions does not match the demand for them.

So fuel oil is **cracked** to meet demand.

CHEMICAL DECOMPOSITION OF COMPOUNDS: CRACKING

The mismatch problem is solved by decomposing some of the fuel oil to make more petrol or diesel. This is called cracking.

The large molecules are broken down or cracked:
- either by heat: **thermal cracking**
- or by catalyst and heat: **catalytic cracking**

Cracking also has the benefit of making reactive, unsaturated compounds with double bonds.

These compounds are important **petrochemical feedstocks.** This means other useful substances can be made from them.

100°C

CRACKER 400°C Al_2O_3

300°C

PETROL AND DIESEL

ETHENE – a petrochemical feedstock

ANTIFREEZE

POLYTHENE

METHYLATED SPIRITS

Typical cracking reaction

$$C_{14}H_{30} \xrightarrow[\text{for between 0.1 and 0.4 seconds}]{750-900°C} C_{12}H_{26} + C_2H_4$$

fuel oil   diesel   ethene

Alkanes

- contain only carbon and hydrogen
- contain only single covalent bonds
- have the maximum amount of hydrogen bonded to the carbon skeleton
- are described as saturated

Name	Molecular formula	Structural formula
methane	CH_4	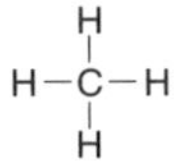 H–C–H (with H above and H below)
ethane	C_2H_6 or CH_3CH_3	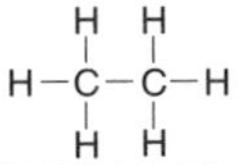 H–C–C–H
propane	C_3H_8 or $CH_3CH_2CH_3$	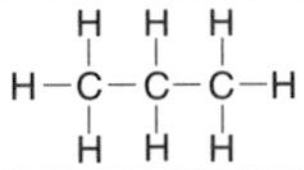 H–C–C–C–H
butane	C_4H_{10} or $CH_3CH_2CH_2CH_3$	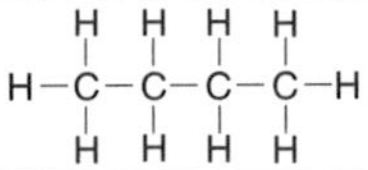 H–C–C–C–C–H
pentane	C_5H_{12} or $CH_3CH_2CH_2CH_2CH_3$	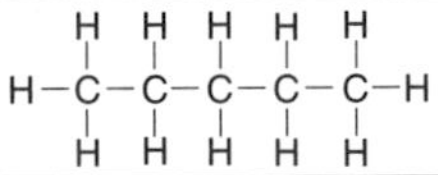 H–C–C–C–C–C–H

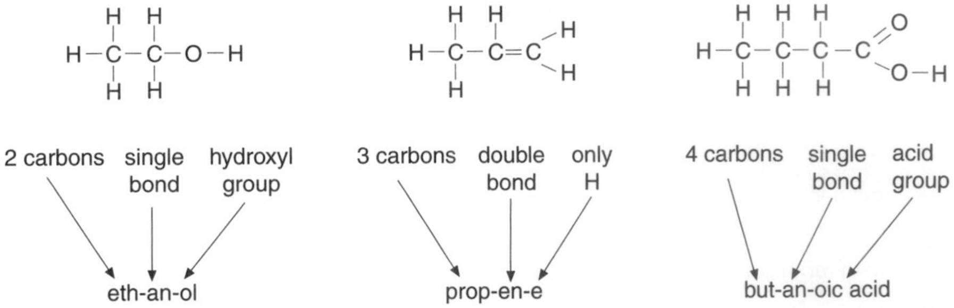

HOMOLOGOUS SERIES

The alkanes are an example of a **homologous series.**

A homologous series is a group of compounds:
1. with the same general formula – the alkanes have the formula C_nH_{2n+2} where n is the number of carbon atoms
2. each of which differs from the next by $–CH_2–$
3. which show a gradual trend in physical properties
4. which have the same chemical reactions.

All the alkanes burn in air.
e.g. $C_3H_8 + 5O_2 \rightarrow 3CO_2 + 4H_2O$

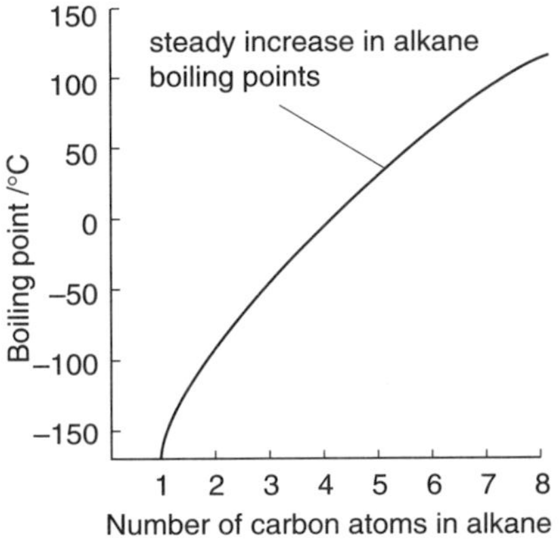

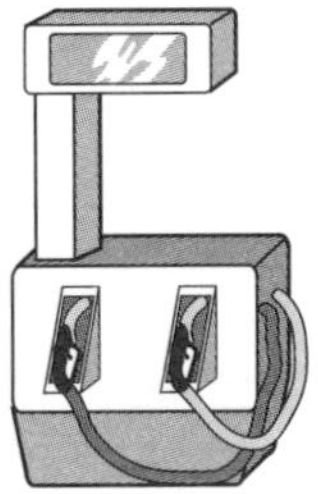

NAMING IN ORGANIC CHEMISTRY

The names of compounds have three parts:

First part: gives the length of the carbon chain
1 carbon → meth-; 2 carbons → eth-; 3 carbons → prop-; 4 carbons → but-

Second part: tells whether there are any double or triple bonds in the carbon chain
all single bonds → -an-; a double bond → -en-; a triple bond -yn-

Third part (ending): tells what is joined to the carbon chain
only hydrogen → -e; a hydroxyl group → -ol; an acid group → -oic acid.

e.g.

2 carbons single bond hydroxyl group → eth-an-ol

3 carbons double bond only H → prop-en-e

4 carbons single bond acid group → but-an-oic acid

ISOMERISM

The atoms of alkanes from butane upwards can be bonded together in more than one way. This is called **isomerism.**

Isomers are molecules with the *same molecular formula but different structural formulas.*
e.g. butane C_4H_{10} can be:

1. a straight chain

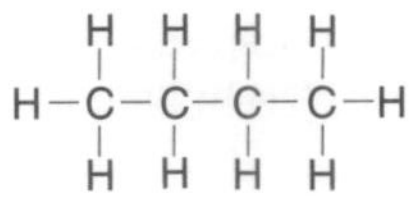

2. a branched chain

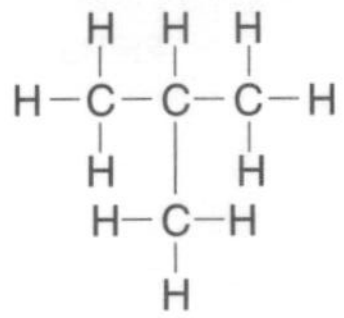

STRUCTURE AND BOILING POINTS

As the graph shows, the longer the chain of carbon atoms, the higher the boiling point. This is because longer chain molecules have bigger forces holding them together.

But the more branched the chain the lower the boiling point, because branched chains cannot pack together so tightly.

steady increase in alkane boiling points

Boiling point /°C — 150, 100, 50, 0, –50, –100, –150

Number of carbon atoms in alkane — 1 2 3 4 5 6 7 8

Alkenes

- contain only carbon and hydrogen
- have a double bond between two of the carbon atoms
- do not have the maximum amount of hydrogen bonded to the carbon skeleton
- are described as unsaturated

$$H_2C=CH_2$$

(structure: $H-C=C-H$ with two H's on each carbon)

Name	Molecular formula	Structural formula
ethene	C_2H_4 or $CH_2=CH_2$	$CH_2=CH_2$
propene	C_3H_6 or $CH_3CH=CH_2$	$CH_3CH=CH_2$ structure
butene	C_4H_8 or $CH_3CH_2CH=CH_2$ or $CH_3CH=CHCH_3$	structures shown and

ANOTHER HOMOLOGOUS SERIES

- general formula C_nH_{2n} where n is the number of carbon atoms
- physical properties show a steady trend as chain length increases
- more stable than alkanes when heated alone (stronger bonding)
- more reactive than alkanes when added to other substances.

OILS AND FATS

Vegetable oils and animal fats are similar compounds but with one important difference. Vegetable oils contain carbon–carbon double bonds: they are **unsaturated** Animal fats contain only carbon–carbon single bonds: they are **saturated**.

Vegetable oils are turned into fats by reacting them with hydrogen. This makes them saturated like animal fats. The process is called **hardening**.

It is cheaper to make fat from vegetable oils in this way than to get fats from animals.

'Hardened' vegetable oils are sold as margarine.

ADDITION REACTIONS

Alkenes combine with bromine, hydrogen, water, and even themselves. These combination reactions are called **additions**.

1. Reaction with bromine

If an alkene is shaken with bromine water, the orange colour of the bromine disappears as the bromine reacts with the alkene.

$$CH_2=CH_2 + Br-Br \rightarrow H-C-C-H \text{ (with Br Br)}$$

colourless   orange   colourless

This reaction is used to distinguish between alkanes and alkenes. If an alkane is shaken with bromine water the colour of the bromine does not go.

2. Reaction with hydrogen

In the presence of a nickel catalyst, hydrogen adds on to an alkene making an alkane.

$$CH_2=CH_2 + H_2 \xrightarrow[180°C]{Ni} CH_3-CH_3$$

This reaction is important for making vegetable oil into fat.

3. Reaction with water

Ethene reacts with water making ethanol (an alcohol).

$$CH_2=CH_2 + H-OH \xrightarrow[\substack{300°C \\ 70 \text{ atmospheres}}]{H_3PO_4 \text{ catalyst}} H-C-C-H \text{ (OH)}$$

4. Reaction with more ethene

Under suitable conditions, ethene molecules react with each other, linking up to form a long chain molecule. This process is called **polymerization**. The ethene reactants are monomers; the long chain product is a polymer.

$$CH_2=CH_2 + CH_2=CH_2 + CH_2=CH_2 \xrightarrow{Ziegler \ catalyst}$$

monomers   part of a polymer chain

Alcohols

HOMOLOGOUS SERIES

Alcohols are a homologous series with the general formula $C_nH_{2n+1}OH$.

The first three members are:

methanol CH_3OH

H—C—O—H
  |
  H

ethanol CH_3CH_2OH

propanol $CH_3CH_2CH_2OH$

Ethanol is the most important alcohol.

INDUSTRIAL MANUFACTURE OF ETHANOL

Ethanol is made by adding water to ethene. This reaction is called **hydration** and is an *addition* reaction.

$$CH_2{=}CH_2(g) + H_2O(g) \xrightarrow[\substack{300°C \\ 70 \text{ atmospheres}}]{H_3PO_4} CH_3CH_2OH(l)$$

This process:
- is continuous
- produces pure ethanol
- uses finite resources (crude oil)
- produces large volumes cheaply

The alcohol produced by this method is 'methylated' by adding methanol. This makes it undrinkable so people cannot drink industrial ethanol and avoid paying duty on the alcohol they drink.

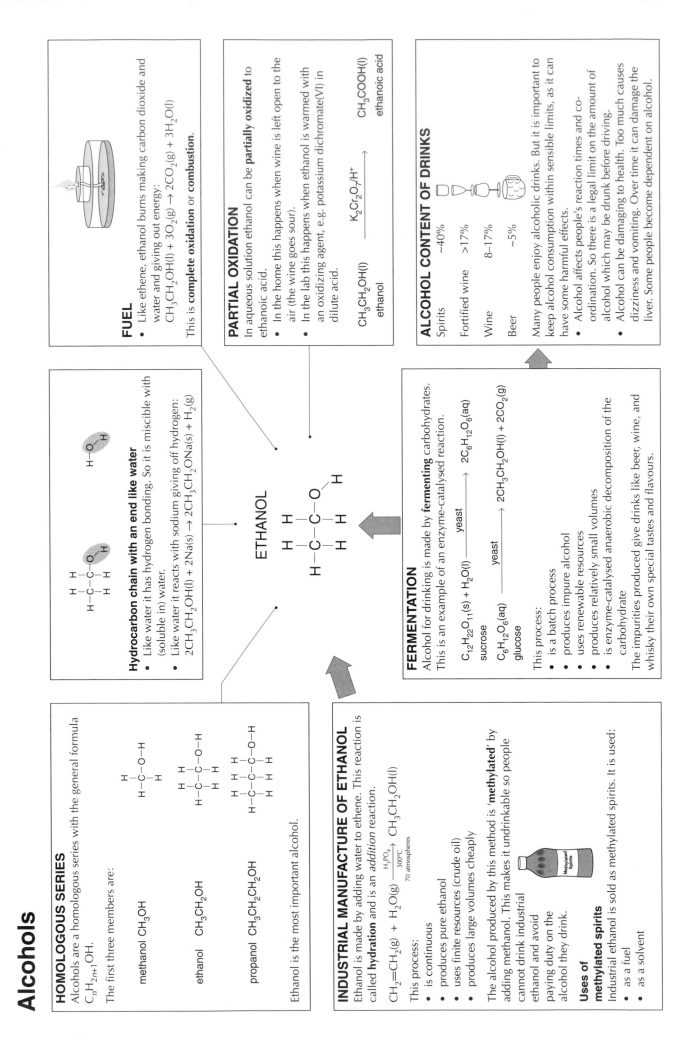

Uses of methylated spirits

Industrial ethanol is sold as methylated spirits. It is used:
- as a fuel
- as a solvent

ETHANOL

H H
| |
H—C—C—O—H
| |
H H

Hydrocarbon chain with an end like water
- Like water it has hydrogen bonding. So it is miscible with (soluble in) water.
- Like water it reacts with sodium giving off hydrogen:
$2CH_3CH_2OH(l) + 2Na(s) \rightarrow 2CH_3CH_2ONa(s) + H_2(g)$

FUEL
- Like ethene, ethanol burns making carbon dioxide and water and giving out energy:
$CH_3CH_2OH(l) + 3O_2(g) \rightarrow 2CO_2(g) + 3H_2O(l)$
This is **complete oxidation** or **combustion**.

PARTIAL OXIDATION

In aqueous solution ethanol can be **partially oxidized** to ethanoic acid.
- In the home this happens when wine is left open to the air (the wine goes sour).
- In the lab this happens when ethanol is warmed with an oxidizing agent, e.g. potassium dichromate(VI) in dilute acid.

$CH_3CH_2OH(l) \xrightarrow{K_2Cr_2O_7/H^+} CH_3COOH(l)$
ethanol   ethanoic acid

FERMENTATION

Alcohol for drinking is made by **fermenting** carbohydrates. This is an example of an enzyme-catalysed reaction.

$$C_{12}H_{22}O_{11}(s) + H_2O(l) \xrightarrow{yeast} 2C_6H_{12}O_6(aq)$$
sucrose

$$C_6H_{12}O_6(aq) \xrightarrow{yeast} 2CH_3CH_2OH(l) + 2CO_2(g)$$
glucose

This process:
- is a batch process
- produces impure alcohol
- uses renewable resources
- produces relatively small volumes
- is enzyme-catalysed anaerobic decomposition of the carbohydrate

The impurities produced give drinks like beer, wine, and whisky their own special tastes and flavours.

ALCOHOL CONTENT OF DRINKS

Spirits	~40%
Fortified wine	>17%
Wine	8–17%
Beer	~5%

Many people enjoy alcoholic drinks. But it is important to keep alcohol consumption within sensible limits, as it can have some harmful effects.
- Alcohol affects people's reaction times and co-ordination. So there is a legal limit on the amount of alcohol which may be drunk before driving.
- Alcohol can be damaging to health. Too much causes dizziness and vomiting. Over time it can damage the liver. Some people become dependent on alcohol.

Carboxylic acids

HOMOLOGOUS SERIES

The carboxylic acids form a homologous series.

The first three are:

methanoic acid   CHOOH

$$\begin{array}{c} O \\ \parallel \\ H-C-O-H \end{array}$$

ethanoic acid CH_3COOH

$$\begin{array}{c} \;\;\;\;\; O \\ \;\;\;\;\; \parallel \\ H-C-C-O-H \\ \;\; | \\ \;\; H \end{array}$$

propanoic acid CH_3CH_2COOH

$$\begin{array}{c} \;\;\;\;\;\;\;\;\;\; O \\ \;\;\;\;\;\;\;\;\;\; \parallel \\ H-C-C-C-O-H \\ \;\; | \;\;\; | \\ \;\; H \;\;\; H \end{array}$$

MAKING ETHANOIC ACID

Ethanoic acid is made by the partial oxidation of ethanol.
Compare this with the complete oxidation.

Partial oxidation

$CH_3CH_2OH(l) + 2[O] \rightarrow CH_3COOH(l) + H_2O(l)$

Complete oxidation

$CH_3CH_2OH(l) + 3O_2(g) \rightarrow 2CO_2(g) + 3H_2O(l)$

Partial oxidation is performed in solution in a test tube or beaker. The oxygen for the oxidation is supplied by acidified potassium dichromate(VI), which is a strong oxidizing agent. The reaction is slow unless it is heated.

Potassium dichromate(VI) is orange. The solution goes *green* during the reaction.

ethanol and potassium dichromate in dilute sulphuric acid

orange colour   changes to   green colour

warm

ethanoic acid

ETHANOIC ACID

$$\begin{array}{c} \;\;\;\; O \\ \;\;\;\; \parallel \\ H-C-C-O-H \\ \;\; | \\ \;\; H \\ \;\; H \end{array}$$

USES

- used in making textiles
- vinegar is impure ethanoic acid.

TYPICAL REACTIONS

1. With water

Carboxylic acids react with water forming hydrogen ions.
They are *weak acids*, so they only form a few ions; most of the molecules remain as molecules.

e.g. $CH_3COOH(l) + H_2O(l) \rightleftharpoons CH_3COO^-(aq) + H^+(aq)$
     99.9% molecules   0.1% ions

So the dissociation of the acid in water is incomplete.

2. With universal indicator

Carboxylic acids change the colour of universal indicator.
They are weak acids so the indicator goes *orange*. (With strong acids the indicator goes red.)

3. With reactive metals

Carboxylic acids react with reactive metals like magnesium:

$Mg(s) + 2CH_3COOH(aq) \rightarrow Mg(CH_3COO)_2(aq) + H_2(g)$

4. With bases

Carboxylic acids are neutralized by bases such as sodium hydroxide or sodium carbonate:

$CH_3COOH(aq) + NaOH(aq) \rightarrow CH_3COONa(aq) + H_2O(l)$

$2CH_3COOH(aq) + Na_2CO_3(aq) \rightarrow 2CH_3COONa(aq) + CO_2(g) - H_2O(l)$

5. With alcohols

Carboxylic acids react with alcohols. A molecule of acid reacts with a molecule of alcohol, splitting out water and making a substance called an **ester**.

$CH_3COOH(l) + HOCH_2CH_3(l) \rightleftharpoons CH_3COOCH_2CH_3(l) + H_2O(l)$
   acid   +   alcohol   ester   +   water

This reaction is *reversible* and *very slow*. It is catalysed by concentrated sulphuric acid.

Esters have fruity smells and flavours. They are used in the food industry.

CONDENSATION POLYMERS

This group of polymers is made by linking up molecules which are reactive at each end, such as amino acids:

$$n\ NH_2 \diagdown\!\!\diagup COOH \rightarrow NH_2 \diagdown\!\!\diagup \underset{O}{\overset{|}{C}} \cdots + n\ H_2O$$

carbon chain

Each time two molecules join together, a small molecule (often water) is also formed as atoms are pushed out to make room for new bonds. This water condenses on the walls of the reaction vessel, so these are called **condensation polymers**.

Nylon and **polyester** are examples of condensation polymers.

ADDITION POLYMERS

Chemists have imitated nature by linking together small molecules to make new materials.

Addition polymers are made from *unsaturated* monomers.

double bond is made into two single bonds.

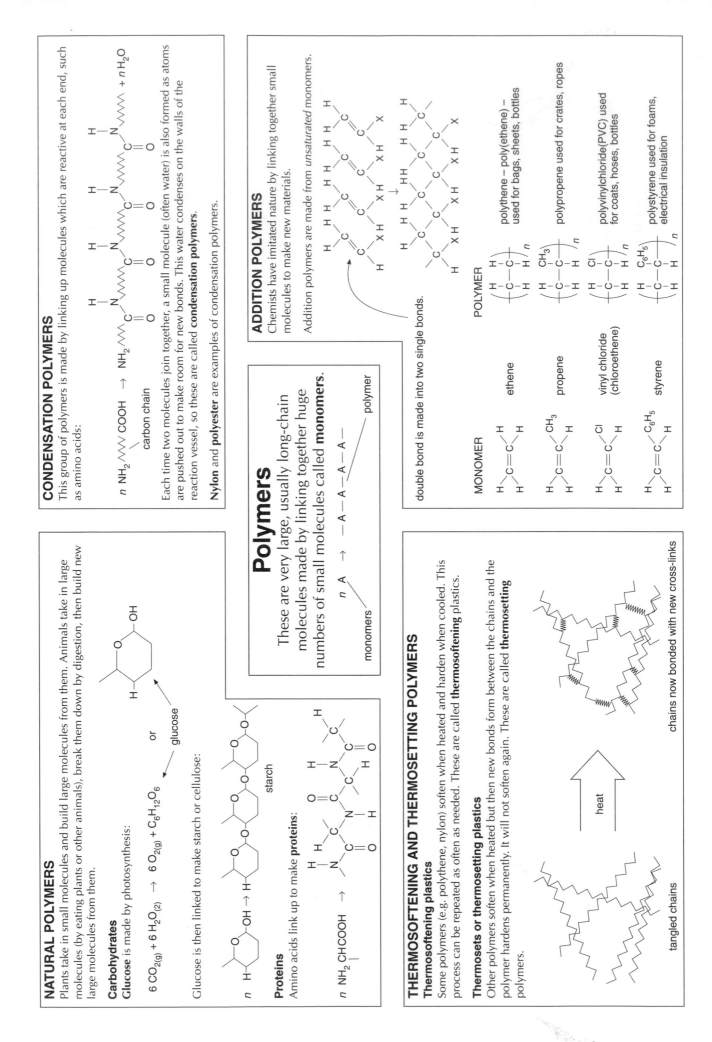

MONOMER		POLYMER	
ethene		polythene – poly(ethene) – used for bags, sheets, bottles	
propene		polypropene used for crates, ropes	
vinyl chloride (chloroethene)		polyvinylchloride(PVC) used for coats, hoses, bottles	
styrene		polystyrene used for foams, electrical insulation	

Polymers

These are very large, usually long-chain molecules made by linking together huge numbers of small molecules called **monomers**.

$$n\ A \rightarrow -A-A-A-A-$$

monomers → polymer

NATURAL POLYMERS

Plants take in small molecules and build large molecules from them. Animals take in large molecules (by eating plants or other animals), break them down by digestion, then build new large molecules from them.

Carbohydrates
Glucose is made by photosynthesis:

$$6\ CO_{2(g)} + 6\ H_2O_{(2)} \rightarrow 6\ O_{2(g)} + C_6H_{12}O_6$$

glucose

Glucose is then linked to make starch or cellulose:

starch

Proteins
Amino acids link up to make **proteins**:

$$n\ NH_2\ CHCOOH \rightarrow$$

THERMOSOFTENING AND THERMOSETTING POLYMERS

Thermosoftening plastics
Some polymers (e.g. polythene, nylon) soften when heated and harden when cooled. This process can be repeated as often as needed. These are called **thermosoftening** plastics.

Thermosets or thermosetting plastics
Other polymers soften when heated but then new bonds form between the chains and the polymer hardens permanently. It will not soften again. These are called **thermosetting** polymers.

tangled chains

heat

chains now bonded with new cross-links

Relative masses and moles

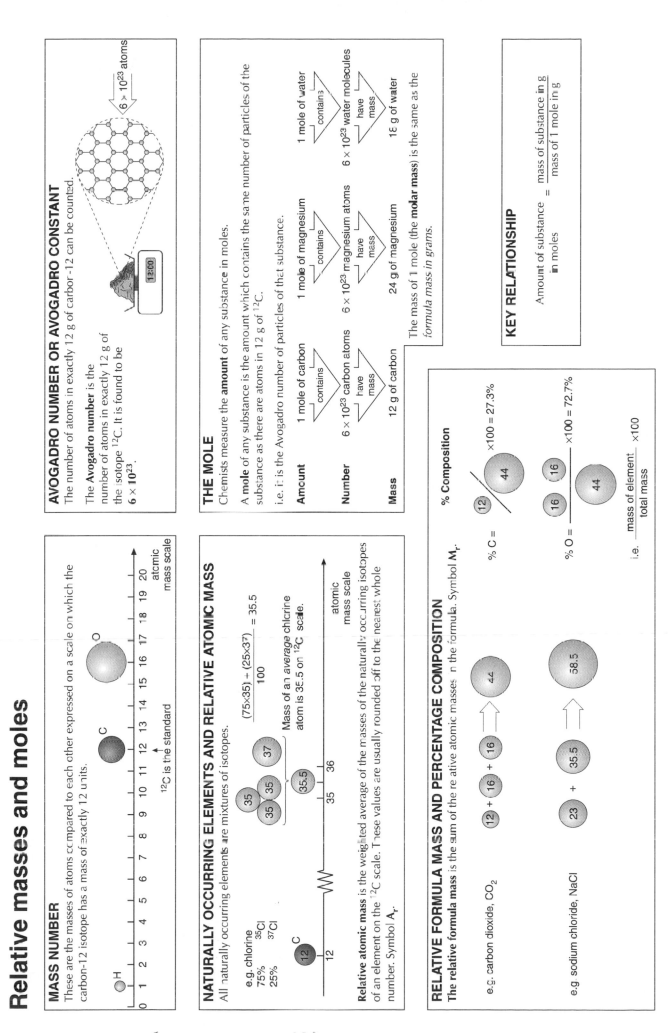

MASS NUMBER

These are the masses of atoms compared to each other expressed on a scale on which the carbon-12 isotope has a mass of exactly 12 units.

^{12}C is the standard

AVOGADRO NUMBER OR AVOGADRO CONSTANT

The number of atoms in exactly 12 g of carbon-12 can be counted.

The **Avogadro number** is the number of atoms in exactly 12 g of the isotope ^{12}C. It is found to be 6×10^{23}.

6×10^{23} atoms

NATURALLY OCCURRING ELEMENTS AND RELATIVE ATOMIC MASS

All naturally occurring elements are mixtures of isotopes.

e.g. chlorine
75% ^{35}Cl
25% ^{37}Cl

$$\frac{(75 \times 35) + (25 \times 37)}{100} = 35.5$$

Mass of an *average* chlorine atom is 35.5 on ^{12}C scale.

Relative atomic mass is the weighted average of the masses of the naturally occurring isotopes of an element on the ^{12}C scale. These values are usually rounded off to the nearest whole number. Symbol A_r.

THE MOLE

Chemists measure the **amount** of any substance in moles.

A **mole** of any substance is the amount which contains the same number of particles of the substance as there are atoms in 12 g of ^{12}C.

i.e. it is the Avogadro number of particles of that substance.

Amount	1 mole of carbon	1 mole of magnesium	1 mole of water
	contains	contains	contains
Number	6×10^{23} carbon atoms	6×10^{23} magnesium atoms	6×10^{23} water molecules
	have mass	have mass	have mass
Mass	12 g of carbon	24 g of magnesium	18 g of water

The mass of 1 mole (the **molar mass**) is the same as the *formula mass in grams*.

RELATIVE FORMULA MASS AND PERCENTAGE COMPOSITION

The **relative formula mass** is the sum of the relative atomic masses in the formula. Symbol M_r.

e.g. carbon dioxide, CO_2

$12 + 16 + 16 \Rightarrow 44$

e.g sodium chloride, NaCl

$23 + 35.5 \Rightarrow 58.5$

% Composition

$$\% C = \frac{12}{44} \times 100 = 27.3\%$$

$$\% O = \frac{16 + 16}{44} \times 100 = 72.7\%$$

i.e. $\dfrac{\text{mass of element}}{\text{total mass}} \times 100$

KEY RELATIONSHIP

$$\text{Amount of substance in moles} = \frac{\text{mass of substance in g}}{\text{mass of 1 mole in g}}$$

Using moles

CHANGING MOLES TO MASS
Key relationship

number of moles × **molar mass** = **mass in g**

N.B. **A mole of oxygen** can mean two things. It can mean
- 1 mole of oxygen atoms which weighs 16 g or
- 1 mole of oxygen molecules which weighs 32 g.

So *always say what particles you are referring to.*

WORKED EXAMPLES
What is the mass of 1.5 moles of magnesium?
$1.5 \times 24 = 36$ g

What is the mass of 0.75 moles of copper(II) oxide?
$0.75 \times (64 + 16) = 0.75 \times 80 = 60$ g

CHANGING MASS TO MOLES
Key relationship

$$\frac{\text{mass in grams}}{\text{molar mass}} = \text{number of moles}$$

WORKED EXAMPLES
How many moles in 10 g of calcium?
$$\frac{10}{40} = 0.25 \text{ mol}$$

How many moles in 10 g of calcium carbonate, $CaCO_3$?
$$\frac{10}{40 + 12 + (16 \times 3)} = \frac{10}{100} = 0.1 \text{ mol}$$

FINDING FORMULAS FROM PERCENTAGE COMPOSITION FIGURES

Use the relationship above to work out the relative number of moles of each element. Then work out the ratio of moles of each element. This formula, which only gives you the *whole number ratio* of atoms is called the **empirical formula**.

To work out the *actual* number of atoms in the molecule (the **molecular formula**), you also have to know the relative formula mass.

Process
1. Assume you have 100g and turn % figures into grams.
2. Convert these masses into moles.
3. Divide these figures by the smallest to get formula.

WORKED EXAMPLES
What is the empirical formula of a hydrocarbon which is 85.7% carbon and 14.3 % hydrogen? Its relative molar mass is 42. What is its molecular formula?

Assume you have 100 g of the hydrocarbon.

The masses are:
carbon 85.7 g   hydrogen 14.3 g

The moles are:
carbon 85.7/12   hydrogen 14.3/1
= 7.14 mol   = 14.3 mol

The simplest ratio of moles is found by dividing by the smallest:
7.14/7.14 = 1   14.3/7.14 = 2

so the empirical formula is C_1H_2.

The formula mass of $C_1H_2 = 12 + 2 = 14$.

The formula mass of the hydrocarbon is 42 which = 3 × empirical formula mass,

so the molecular formula is C_3H_6.

A compound has the percentage composition: sodium 29.1%; sulphur 40.5 %; oxygen 30.4%. Its relative formula mass is 158. What is its formula?

	Na	S	O
% composition	29.1	40.5	30.4

Assume you have 100 g of the compound.

The masses are:

	29.1 g	40.5 g	30.4 g

The moles are:

	29.1/23	40.5/32	30.4/16
	= 1.27 mol	= 1.27 mol	= 1.9 mol

Divided by the smaller number :

1.27/1.27	:	1.27/1.27	:	1.9/1.27
= 1	:	1	:	1.5

There must be whole numbers of atoms, so this becomes

2	:	2	:	3

The empirical formula is $Na_2S_2O_3$. This has a formula mass of 158 so the molecular formula is also $Na_2S_2O_3$.

Moles and concentrations of solutions

Concentration is a measure of how much solute is dissolved in a solvent.

Concentrations are measured in **moles per unit volume**. The unit of volume used may be the dm³ or the litre. This is not as confusing as it seems because 1 dm³ = 1 litre.

So **concentration = mol / dm³** (sometimes written mol dm⁻³) or **mol / litre** (sometimes written mol l⁻¹).

Concentration in mol / litre is sometimes known as **molarity** (written as M).

It is important to remember that:

$1 \ dm^3 = 1000 \ cm^3 = 1 \ litre = 1000 \ ml$ which means that

concentration = mol / dm³ = mol × 1000 / volume in cm³

To calculate the amount of solute in solution, you need to know both the concentration and volume of the solution.

MASS AND VOLUME TO CONCENTRATION
Process
1. Calculate moles of solute
2. Scale volume to 1 dm³

WORKED EXAMPLES

5.85 g of sodium chloride is dissolved in 200 cm³ of water. Calculate the concentration in mol / dm³.

5.85 g of sodium chloride is 5.85 / (23 + 35.5) = 0.1 mol

so logically:

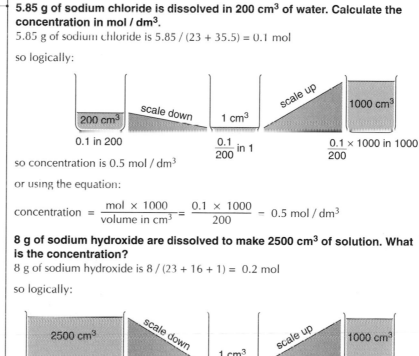

0.1 in 200 $\frac{0.1}{200}$ in 1 0.1×1000 in 1000 / 200

so concentration is 0.5 mol / dm³

or using the equation:

$$\text{concentration} = \frac{mol \times 1000}{\text{volume in } cm^3} = \frac{0.1 \times 1000}{200} = 0.5 \ mol / dm^3$$

8 g of sodium hydroxide are dissolved to make 2500 cm³ of solution. What is the concentration?

8 g of sodium hydroxide is 8 / (23 + 16 + 1) = 0.2 mol

so logically:

0.2 mol in 2500 $\frac{0.2}{2500}$ in 1 0.2×1000 in 1000 / 2500

so concentration is 0.08 mol / dm³

or using the equation:

$$\text{concentration} = \frac{mol \times 1000}{\text{volume in } cm^3} = \frac{0.2 \times 1000}{2500} = 0.08 \ mol / dm^3$$

CONCENTRATION AND VOLUME TO MASS
Process
1. Write down the number of moles in 1 dm³
2. Scale moles to the volume in the question
3. Convert from moles to mass

WORKED EXAMPLE
What mass of ammonium chloride is dissolved in 250 cm³ of a 0.1 molar (0.1 mol / dm³) solution?

In 1 dm³ there is 0.1 mol

so logically:

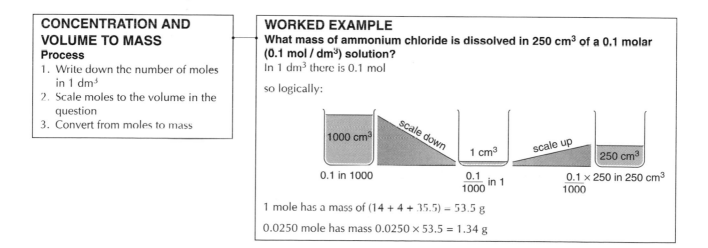

0.1 in 1000 $\frac{0.1}{1000}$ in 1 0.1×250 in 250 cm³ / 1000

1 mole has a mass of (14 + 4 + 35.5) = 53.5 g

0.0250 mole has mass 0.0250 × 53.5 = 1.34 g

Moles and volumes of gases

MOLAR VOLUME

Experiments show that **1 mole of any gas has a volume of 24 dm³** (24 000 cm³) **at room temperature and atmospheric pressure** (r.t.p.).

This is an application of **Avogadro's Law** which states that **equal volumes of gases under the same conditions contain equal numbers of particles**.

At **standard temperature and pressure** (s.t.p.) 1 mole of a gas has a volume of 22.4 dm³.

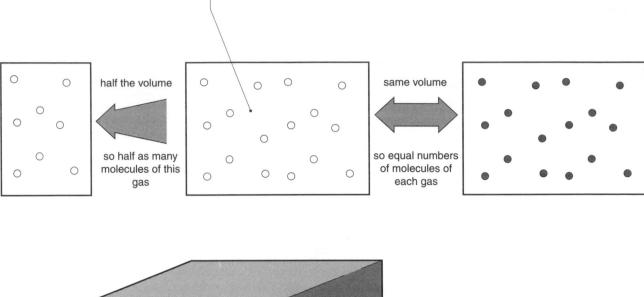

half the volume

so half as many molecules of this gas

same volume

so equal numbers of molecules of each gas

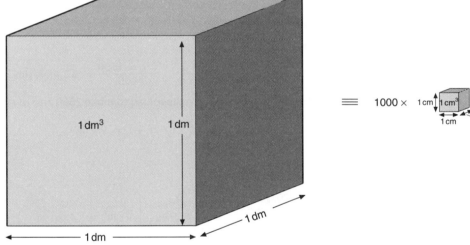

1 dm³   1 dm

1 dm

1 dm

$\equiv$ 1000 ×   1 cm 1 cm³

1 cm

1 cm

CHANGING VOLUMES TO MOLES
Key relationship

$$\frac{\text{Volume of the gas in dm}^3}{24} = \text{number of moles of the gas}$$

$$\frac{\text{Volume of gas in cm}^3}{24\ 000} = \text{number of moles of the gas}$$

WORKED EXAMPLES
How many moles are there in 0.08 dm³ of hydrogen at r.t.p.?
0.08/24 = 0.00333 mol

How many moles in 125 cm³ of hydrogen chloride at r.t.p.?
125/24 000 = 0.00521 mol

CHANGING MOLES TO VOLUMES
Key relationship

Number of moles × 24 = volume of gas in dm³

Number of moles × 24 000 = volume of gas in cm³

WORKED EXAMPLES
What is the volume of 1.5 moles of air at r.t.p.?
1.5 × 24 = 36 dm³

What is the volume of 0.05 moles of chlorine at r.t.p.?
0.05 × 24 000 = 1200 cm³ or 1.2 dm³

Calculations from equations 1

In a **balanced equation** the mass of the reactants always equals the mass of the products. The equation shows matter being rearranged. Matter cannot be destroyed, so the mass of the products must be the same as the mass of the reactants.

An equation can be interpreted in two ways. For example, the equation:
$2H_2(g) + O_2(g) \rightarrow 2H_2O(l)$

- can be seen as representing the particles which take part in the reaction:
  2 hydrogen molecules react with **1 oxygen molecule** to make **2 water molecules**

- or can be seen as relating the number of moles which have reacted:
  2 moles of hydrogen react with **1 mole** of oxygen to make **2 moles** of water.

Of course one equation is just a scaled up version of the other, the scale factor being the Avogadro number.

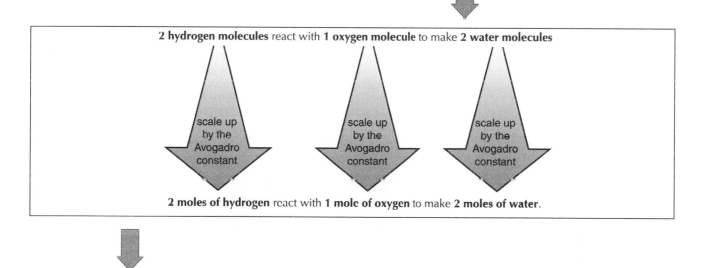

2 hydrogen molecules react with **1 oxygen molecule** to make **2 water molecules**

scale up by the Avogadro constant

scale up by the Avogadro constant

scale up by the Avogadro constant

2 moles of hydrogen react with **1 mole of oxygen** to make **2 moles of water**.

In the laboratory we think of equations in terms of moles, because we can turn moles into masses of solids (which we can weigh out) or into volumes of gases (which we can measure with a gas syringe).

So equations can be used to calculate reacting quantities.

PROCESS
1. Write out the balanced equation
2. Write out the mole ratios
3. Work out the moles of the known or given substance
4. Rewrite the mole ratio using the molar amount found in 3 above
5. Work out the moles of the unknown or wanted substance

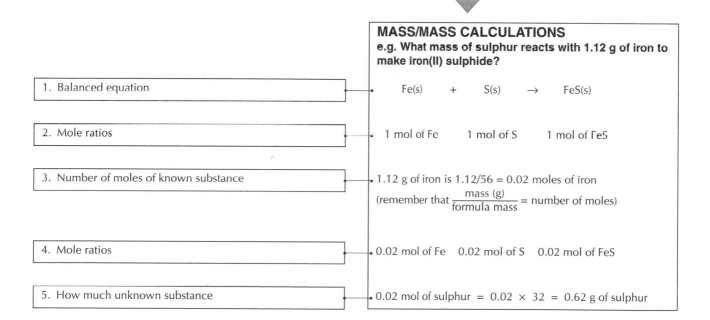

MASS/MASS CALCULATIONS
e.g. What mass of sulphur reacts with 1.12 g of iron to make iron(II) sulphide?

1. Balanced equation	$Fe(s) + S(s) \rightarrow FeS(s)$
2. Mole ratios	1 mol of Fe 1 mol of S 1 mol of FeS
3. Number of moles of known substance	1.12 g of iron is $1.12/56 = 0.02$ moles of iron (remember that $\dfrac{\text{mass (g)}}{\text{formula mass}}$ = number of moles)
4. Mole ratios	0.02 mol of Fe 0.02 mol of S 0.02 mol of FeS
5. How much unknown substance	0.02 mol of sulphur $= 0.02 \times 32 = 0.62$ g of sulphur

Calculations from equations 2

MASS/MASS CALCULATIONS
e.g. What mass of aluminium powder will burn in 1.6 g of oxygen?

1. Balanced equation

$$4Al(s) \quad + \quad 3O_2(g) \quad \rightarrow \quad 2Al_2O_3(s)$$

2. Mole ratios

4 mol of Al   3 mol of O_2 2 mol of Al_2O_3

3. Number of moles of known substance

1.6 g of oxygen is $\dfrac{1.6}{32}$ = 0.05 mol

(remember that $\dfrac{\text{mass (g)}}{\text{formula mass}}$ = number of moles)

4. Mole ratios

From step 2:

	4 mol	3 mol	2 mol
	4 mol	3 mol	2 mol
so	4/3 mol	1 mol	2/3 mol
so	4/3 × 0.05 mol	0.05 mol	2/3 × 0.05 mol

5. How much unknown substance

4/3 × 0.05 mol of aluminium is 0.067 mol. This has a mass of 0.067 × 27 = 1.8 g.

MASS/VOLUME CALCULATIONS
e.g. What volume of ammonia reacts with 4 g of copper oxide?

1. Balanced equation

$$3CuO(s) \quad + \quad 2NH_3(g) \quad \rightarrow \quad 3Cu(s) \quad + \quad 3H_2O(g) \quad + \quad N_2(g)$$

2. Mole ratios

3 mol of CuO   2 mol of NH_3 3 mol of Cu   3 mol of H_2O 1 mol of N_2

3. Number of moles of known substance

4 g of copper oxide is $\dfrac{4}{(64 + 16)}$ = 0.05 mol

(remember that $\dfrac{\text{mass (g)}}{\text{formula mass}}$ = number of moles)

4. Mole ratios

Step 2 gave

	3 mol	2 mol	3 mol	3 mol	1 mol
	3 mol	2 mol	3 mol	3 mol	1 mol
so	1 mol	2/3 mol	1 mol	1 mol	1/3 mol
so	0.05 mol	2/3 × 0.05 mol	0.05 mol	0.05 mol	1/3 × 0.05 mol

5. How much unknown substance

2/3 × 0.05 mol of ammonia has a volume of 2/3 × 0.05 × 24000 = 800 cm³

VOLUME/VOLUME CALCULATIONS FOR GASES
e.g. What volume of oxygen is needed for the complete combustion of 50 cm³ of propane, C₃H₈?

1. Balanced equation

$$C_3H_8(g) \quad + \quad 5O_2(g) \quad \rightarrow \quad 3CO_2(g) \quad + \quad 4H_2O(g)$$

2. Mole ratios

1 mol of C_3H_8 5 mol of O_2 3 mol of CO_2 4 mol of H_2O

3. Now remember Avogadro's law. Equal volumes of gases contain equal numbers of particles. That means that equal numbers of molecules occupy equal volumes. So we can simply convert from a mole ratio to a volume ratio.

1 volume   5 volumes   3 volumes   4 volumes

and so 50 cm³ of C_3H_8 5 × 50 cm³ of O_2 3 × 50 cm³ of CO_2 4 × 50 cm³ of H_2O

so the volume of oxygen needed is 250 cm³.

Calculations from equations 3

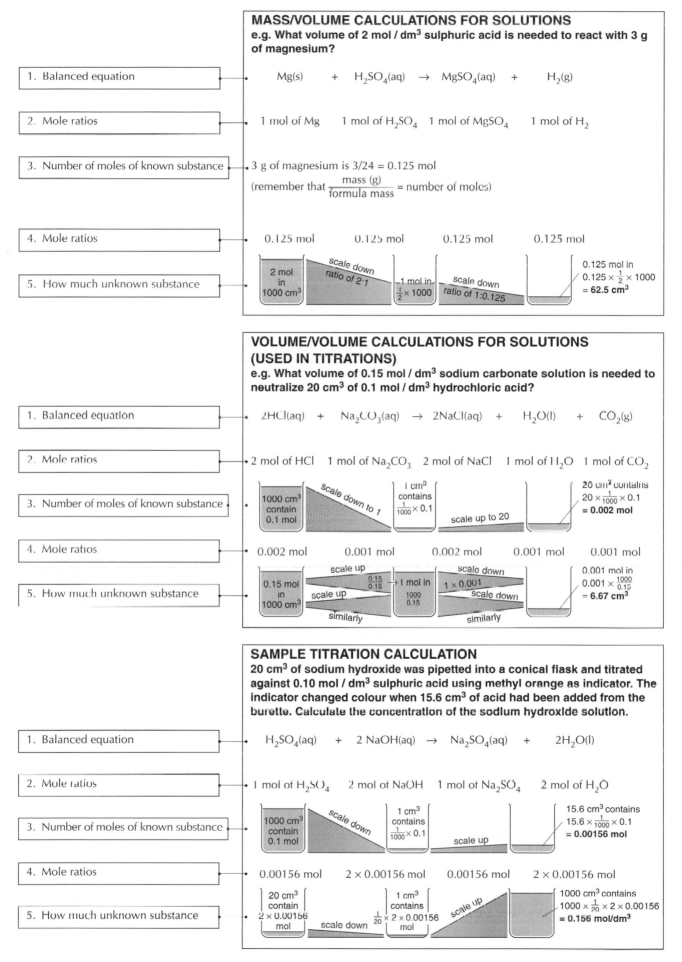

MASS/VOLUME CALCULATIONS FOR SOLUTIONS

e.g. What volume of 2 mol / dm³ sulphuric acid is needed to react with 3 g of magnesium?

1. Balanced equation

$$Mg(s) \quad + \quad H_2SO_4(aq) \quad \rightarrow \quad MgSO_4(aq) \quad + \quad H_2(g)$$

2. Mole ratios

1 mol of Mg   1 mol of H_2SO_4 1 mol of $MgSO_4$ 1 mol of H_2

3. Number of moles of known substance

3 g of magnesium is 3/24 = 0.125 mol

(remember that $\dfrac{\text{mass (g)}}{\text{formula mass}}$ = number of moles)

4. Mole ratios

0.125 mol   0.125 mol   0.125 mol   0.125 mol

5. How much unknown substance

2 mol in 1000 cm³ — scale down ratio of 2:1 → 1 mol in $\frac{1}{2}$ × 1000 — scale down ratio of 1:0.125 → 0.125 mol in 0.125 × $\frac{1}{2}$ × 1000 = **62.5 cm³**

VOLUME/VOLUME CALCULATIONS FOR SOLUTIONS (USED IN TITRATIONS)

e.g. What volume of 0.15 mol / dm³ sodium carbonate solution is needed to neutralize 20 cm³ of 0.1 mol / dm³ hydrochloric acid?

1. Balanced equation

$$2HCl(aq) \quad + \quad Na_2CO_3(aq) \quad \rightarrow \quad 2NaCl(aq) \quad + \quad H_2O(l) \quad + \quad CO_2(g)$$

2. Mole ratios

2 mol of HCl   1 mol of Na_2CO_3 2 mol of NaCl   1 mol of H_2O 1 mol of CO_2

3. Number of moles of known substance

1000 cm³ contain 0.1 mol — scale down to 1 → 1 cm³ contains $\frac{1}{1000}$ × 0.1 — scale up to 20 → 20 cm³ contains 20 × $\frac{1}{1000}$ × 0.1 = **0.002 mol**

4. Mole ratios

0.002 mol   0.001 mol   0.002 mol   0.001 mol   0.001 mol

5. How much unknown substance

0.15 mol in 1000 cm³ — scale up $\frac{0.15}{0.15}$ → 1 mol in $\frac{1000}{0.15}$ — scale down 1 × 0.001 → 0.001 mol in 0.001 × $\frac{1000}{0.15}$ = **6.67 cm³** (similarly)

SAMPLE TITRATION CALCULATION

20 cm³ of sodium hydroxide was pipetted into a conical flask and titrated against 0.10 mol / dm³ sulphuric acid using methyl orange as indicator. The indicator changed colour when 15.6 cm³ of acid had been added from the burette. Calculate the concentration of the sodium hydroxide solution.

1. Balanced equation

$$H_2SO_4(aq) \quad + \quad 2NaOH(aq) \quad \rightarrow \quad Na_2SO_4(aq) \quad + \quad 2H_2O(l)$$

2. Mole ratios

1 mol of H_2SO_4 2 mol of NaOH   1 mol of Na_2SO_4 2 mol of H_2O

3. Number of moles of known substance

1000 cm³ contain 0.1 mol — scale down → 1 cm³ contains $\frac{1}{1000}$ × 0.1 — scale up → 15.6 cm³ contains 15.6 × $\frac{1}{1000}$ × 0.1 = **0.00156 mol**

4. Mole ratios

0.00156 mol   2 × 0.00156 mol   0.00156 mol   2 × 0.00156 mol

5. How much unknown substance

20 cm³ contain 2 × 0.00156 mol — scale down $\frac{1}{20}$ → 1 cm³ contains $\frac{1}{20}$ × 2 × 0.00156 mol — scale up → 1000 cm³ contains 1000 × $\frac{1}{20}$ × 2 × 0.00156 = **0.156 mol/dm³**

Electrolysis calculations

Electrons coming from anode

Anions lose electrons. Anode reactions are oxidations.

Electrons going to the cathode

Cations gain electrons. Cathode reactions are reductions.

$M^{2+}(l)$

$M(s)$

$X_2(g)$

ELECTROLYSIS

In electrolysis electrons are lost and gained by ions at the electrodes. These electrons are pumped round the circuit by a battery or power pack.

The amount of electrolysis depends on the number of electrons which flow round the circuit, i.e. the amount of **charge**. Charge is measured in **coulombs**. The relationship between current, time, and charge is:

amps $\times$ seconds = coulombs

The charge carried by **1 mole of electrons** is called the **Faraday**.

1 Faraday is 96 500 coulombs.

The mass of substance discharged at an electrode depends on:

- **the current flowing**
- **the time the current flows for**
- **the charge on the ion**
  Ions with a double charge need twice as many electrons to discharge them as ions with a single charge.
  $Ca^{2+} + 2e^- \rightarrow Ca$ compared with $Na^+ + 1e^- \rightarrow Na$

ELECTROLYSIS CALCULATIONS

Follow the same five steps as in other calculations from equations.

e.g. A current of 12 amps flows through molten lead bromide for 15 minutes. Calculate the masses of lead and bromine produced at the electrodes.

1. Balanced equations

Anode: $2Br^-(l) \rightarrow Br_2(g) + 2e^-$ **Cathode**: $Pb^{2+}(l) + 2e^- \rightarrow Pb(l)$

2. Mole ratios

2 mol   1 mol   2 mol   1 mol   2 mol   1 mol

3. Number of moles of electrons

charge $= 12 \times 15 \times 60$ (current $\times$ time *in seconds*)
    $= 10\ 800$ coulombs
which is $10\ 800/96\ 500 = 0.112$ mol of electrons (Faraday)

4. Mole ratios

0.112 0.5×0.112 0.112 0.5×0.112 0.112 0.5×0.112

5. How much unknown substance

0.5×0.112 mol of bromine $= 0.5 \times 0.112 \times 79 \times 2 = 8.84$ g of bromine
0.5×0.112 mol of lead $= 0.5 \times 0.112 \times 208 = 11.6$ g of lead

e.g. A current of 100 000 amps is passed through the Hall cell in an aluminium plant for 1 hour. What mass of aluminium is deposited and what volume of oxygen is produced?

1. Balanced equations

Anode: $2O^{2-}(l) \rightarrow O_2(g) + 4e^-$ **Cathode**: $Al^{3+}(l) + 3e^- \rightarrow Al(l)$

2. Mole ratios

2 mol   1 mol   4 mol   1 mol   3 mol   1 mol

3. Number of moles of electrons

charge $= 100\ 000 \times 60 \times 60$ (current x time in seconds)
    $= 360\ 000\ 000$ coulombs
which is $360\ 000\ 000/96\ 500 = 3730$ mol of electrons

4. Mole ratios

$1/2 \times 3730$ $1/4 \times 3730$ 3730 $1/3 \times 3730$ 3730 $1/3 \times 3730$

5. How much unknown substance

$1/4 \times 3730$ mol of oxygen $= 1/4 \times 3730 \times 24$ dm^3
    $= 21\ 500$ dm^3
$1/3 \times 3730$ mol of aluminium $= 1/3 \times 3730 \times 27$ g
    $= 33\ 200$ g
    $= 33.2$ kg

Making salts in the laboratory

SOLUBILITY RULES FOR SALTS
- all Group 1 compounds are soluble
- all nitrates are soluble
- all chlorides are soluble except silver and lead
- all sulphates are soluble except calcium, barium, and lead
- all carbonates are insoluble except Group 1

SOLUBILITY RULES FOR BASES
- all Group 1 bases are soluble
- in Group 2, calcium and barium salts are soluble
- ammonia is soluble

DECIDE IF THE SALT IS SOLUBLE USING SOLUBILITY RULES

Salt is soluble

Salt is insoluble

If you are adding a soluble base

USE TITRATION

1. Measure out a volume of alkaline solution, say 20 cm³. Add some phenolphthalein indicator, then run acid into the flask from a burette.
2. Continue adding acid until one more drop of acid is just enough to turn the solution from pink to colourless. The base has now been neutralized.
3. Note the volume of acid needed for the neutralization this is the reading on the burette.
4. Repeat the process, this time without adding indicator. Run the measured volume of acid from the burette into the 20 cm³ of alkali in the flask.
5. Evaporate off most of the water from the solution over a water bath. Leave the remaining solution to cool, so that the salt can crystallize out. Dry off the crystals on filter paper

Burette
Acid
Alkali
Start
Finish
21
22
0
1
Salt solution
Boiling water
Heat

MAKE THE SALT BY NEUTRALIZATION
- Choose the appropriate acid, e.g. HCl for chlorides, HNO₃ for nitrates, H₂SO₄ for sulphates
- add either
  (a) a reactive metal: Mg, Al, Fe, Zn or (b) a base: a metal oxide, hydroxide, or carbonate
- if you are adding a base, decide whether it is soluble

If you are adding a metal or insoluble base

USE EXCESS BASE METHOD

1. Measure out a volume of acid solution. Add a spatula full of solid: metal, metal oxide, metal hydroxide, or metal carbonate. Warm the beaker.
2. Continue adding solid reactant to the solution until no more dissolves. Warm and stir the solution to make sure the reaction is complete.
3. Filter off the excess solid and collect the filtrate, which is a pure salt solution.
4. Evaporate off most of the water from the solution over a warm bath. Leave the remaining solution to cool, so that the salt can crystallize out. Dry off the crystals on filter paper.

Base
Acid solution
Neutralized acid
Excess solid
Pure salt solution
Salt solution
Boiling water
Heat

MAKE THE SALT BY PRECIPITATION

Use the solubility rules to think of two solutions, one having the cation, the other the anion.

1. Add the cation solution to the anion solution.

Cation solution
Precipitate of the insoluble salt forming in anion solution

2. Filter the resulting suspension.

Suspension of insoluble salt
Insoluble salt forms residue

3. Wash then dry the residue.

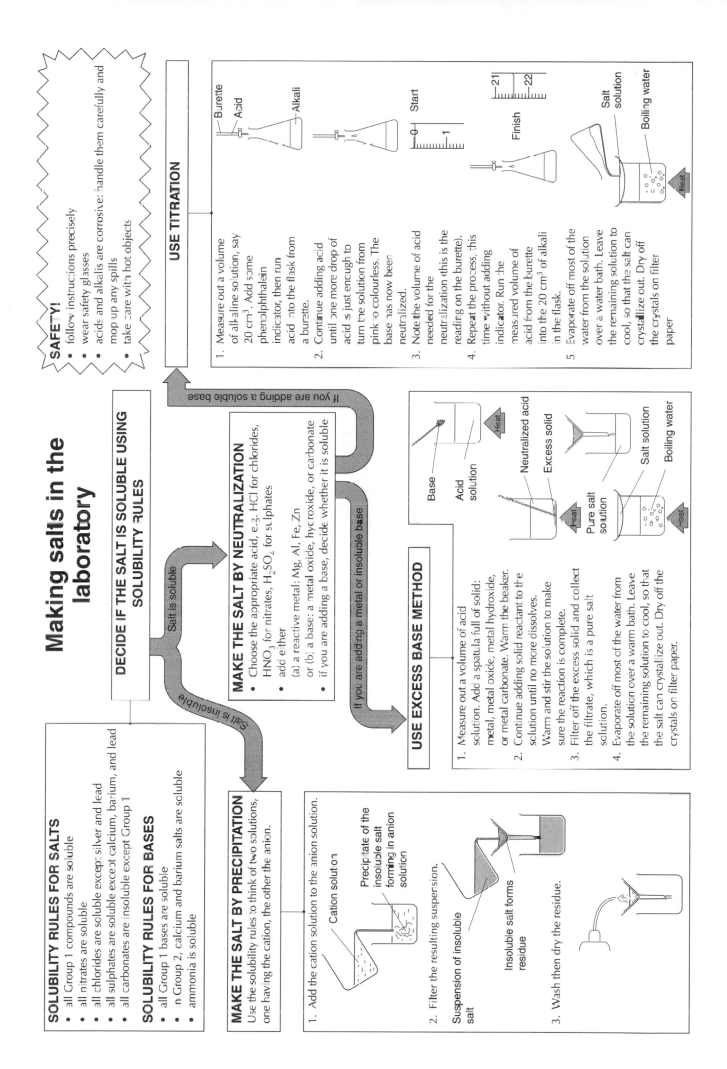

Measuring solubility

It is easier to see the first crystals forming as a saturated solution cools than it is to judge when the very last crystal dissolves as the solution is heated. So to measure solubility, you cool solutions of different concentrations and record the temperature when the first crystals appear on the surface (the coolest part of the solution) and then sink to the bottom. If you miss this moment, you can always reheat the solution and repeat the reading.

The different solutions are made by adding more and more water to the same mass of solute.

PROCEDURE FOR POTASSIUM NITRATE

1. Weigh a boiling tube.

    _23.9_

2. Add about 12 g of potassium nitrate, and re-weigh.

    _35.8_

3. Add 10 cm³ (10 g) of water from a burette.

    Burette
    Water

4. Heat until all the crystals of potassium nitrate have dissolved.

5. Clamp a thermometer so that it hangs in the middle of the tube.

6. Record the temperature at which crystals first appear and sink.

    _T_ °C

7. Add 2 cm³ (2 g) more water to the solution in the tube.

8. Repeat steps 4, 5, and 6.

9. Repeat steps 7 and 8 until 20 cm³ (20 g) water in all have been added.

10. Tabulate readings and calculate solubility in g / 100 g for each temperature.

11. Plot the results to give a solubility curve.

Solubility (in g per 100g of water)

Temperature (in °C)

Titrations

PROCEDURE

In a titration you react a known volume of one solution (measured in a **pipette** with a known volume of another solution (measured in a **burette**). You can tell when the reaction is complete because

- *either* an indicator which has been added changes colour, *or*
- one of the solutions itself changes colour.

SAFETY!
- follow instructions precisely
- wear safety glasses
- acids and alkalis are corrosive: handle them carefully and mop up any spills

1. Label and fill two beakers with the reacting solutions.

ACID   ALKALI

2. Pour a little of one solution into the burette to rinse it. Check the tap works. Pour away all this solution.

3. Fill the burette. Run liquid through the tap until there are no air bubbles. Zero the burette.

4. Using a pipette filler, rinse out the pipette with the other solution. Then refill the pipette to well above the mark.

pipette filler

5. Carefully let the solution out until the meniscus is on the line.

6. Run this solution into a conical flask. Do not force out the last drop, but just touch the pipette tip on the liquid surface.

7. Add just enough indicator to produce a definite colour (about 2–5 drops). Put the conical flask on a white tile. Adjust the height of the burette so that it just sticks into the neck of the conical flask.

8. Run solution from the burette into the conical flask, swirling the flask all the time. Do not shake the flask, or solution will get splashed up the sides where it may not react.

9. Stop when the indicator changes colour and note the volume run in from the burette.

10. Empty and rinse the conical flask. It need not be dried because any water will dilute both solutions equally.

11. Refill and re-zero the burette.

12. Using the pipette filler, refill the pipette and zero it.

13. Repeat steps 5, 6, and 7.

14. Repeat step 3 but this time stop about 1 cm³ before the previously recorded volume. Then add the solution drop by drop until the colour just changes.

15. Record the volume added in a table

Final reading
Initial reading
Volume added

16. Do at least two accurate titrations and work out the average volume added.

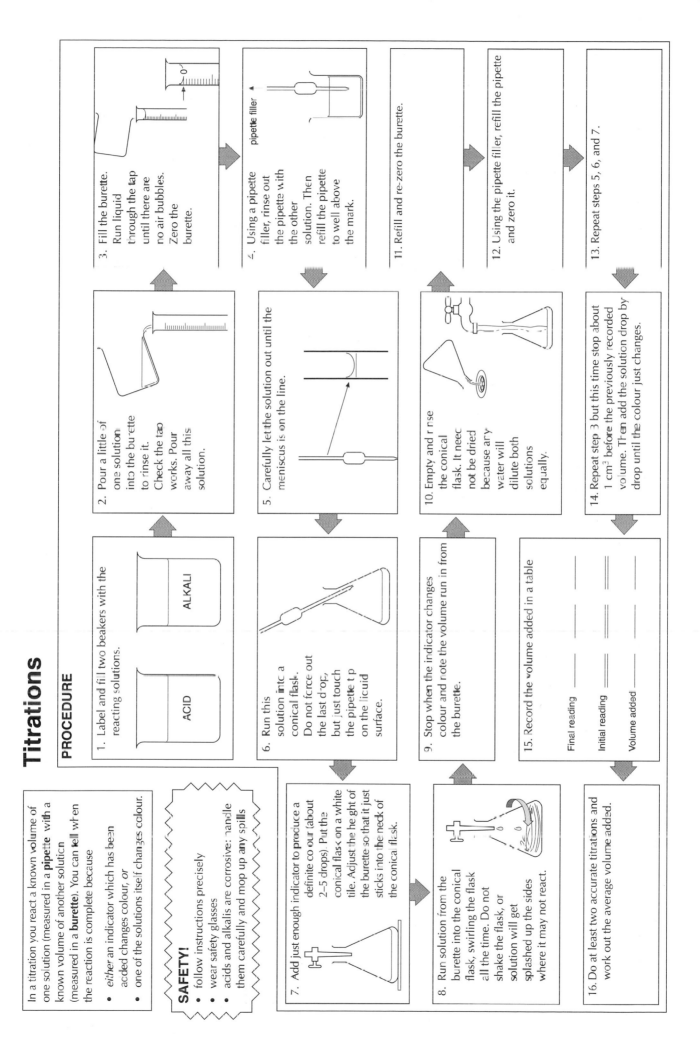

Testing for cations

SAFETY!
- follow instructions precisely
- wear safety glasses
- acids and alkalis are corrosive: handle them carefully and mop up any spills
- take care with hot objects

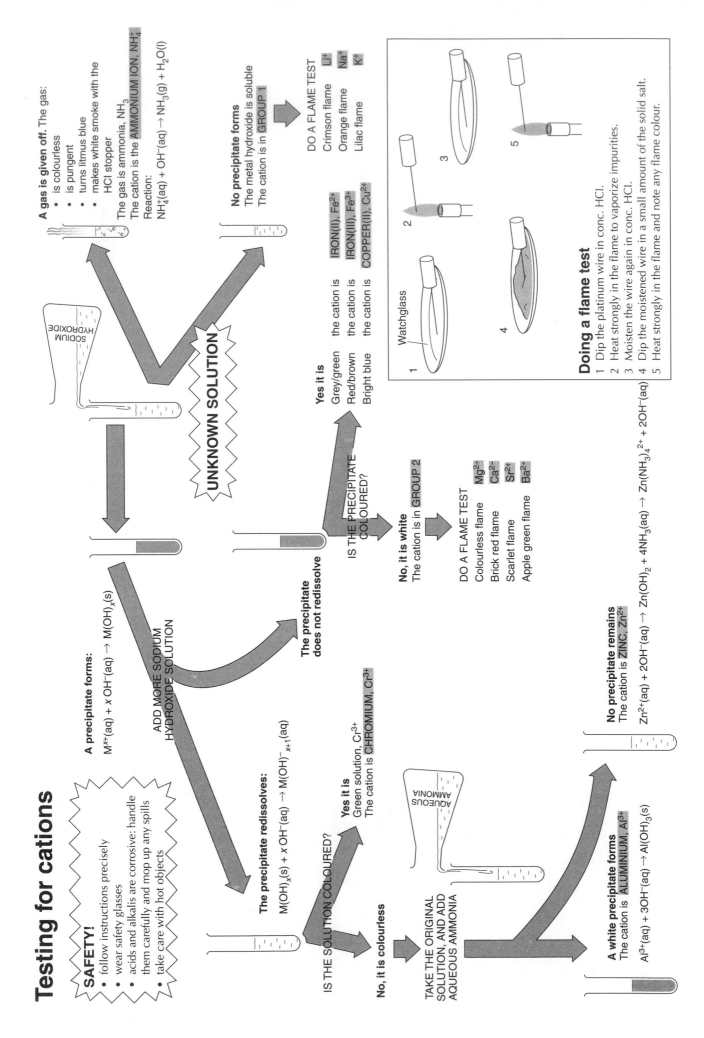

UNKNOWN SOLUTION

SODIUM HYDROXIDE

A gas is given off. The gas:
- is colourless
- is pungent
- turns litmus blue
- makes white smoke with the HCl stopper

The gas is ammonia, NH_3
The cation is the **AMMONIUM ION**, NH_4^+
Reaction:
$NH_4^+(aq) + OH^-(aq) \rightarrow NH_3(g) + H_2O(l)$

No precipitate forms
The metal hydroxide is soluble
The cation is in GROUP 1

DO A FLAME TEST
Crimson flame Li^+
Orange flame Na^+
Lilac flame K^+

A precipitate forms:
$M^{x+}(aq) + x\,OH^-(aq) \rightarrow M(OH)_x(s)$

ADD MORE SODIUM HYDROXIDE SOLUTION

The precipitate redissolves:
$M(OH)_x(s) + x\,OH^-(aq) \rightarrow M(OH)^-_{x+1}(aq)$

IS THE SOLUTION COLOURED?

Yes it is
Green solution, Cr^{3+}
The cation is CHROMIUM, Cr^{3+}

No, it is colourless

TAKE THE ORIGINAL SOLUTION, AND ADD AQUEOUS AMMONIA

AQUEOUS AMMONIA

A white precipitate forms
The cation is ALUMINIUM, Al^{3+}
$Al^{3+}(aq) + 3OH^-(aq) \rightarrow Al(OH)_3(s)$

No precipitate remains
The cation is ZINC, Zn^{2+}
$Zn^{2+}(aq) + 2OH^-(aq) \rightarrow Zn(OH)_2(s) + 4NH_3(aq) \rightarrow Zn(NH_3)_4^{2+} + 2OH^-(aq)$

The precipitate does not redissolve

IS THE PRECIPITATE COLOURED?

Yes it is
Grey/green   the cation is   IRON(II), Fe^{2+}
Red/brown   the cation is   IRON(III), Fe^{3+}
Bright blue the cation is   COPPER(II), Cu^{2+}

No, it is white
The cation is in GROUP 2

DO A FLAME TEST
Colourless flame Mg^{2+}
Brick red flame Ca^{2+}
Scarlet flame Sr^{2+}
Apple green flame Ba^{2+}

Doing a flame test
1 Dip the platinum wire in conc. HCl.
2 Heat strongly in the flame to vaporize impurities.
3 Moisten the wire again in conc. HCl.
4 Dip the moistened wire in a small amount of the solid salt.
5 Heat strongly in the flame and note any flame colour.

Watchglass

Testing for anions

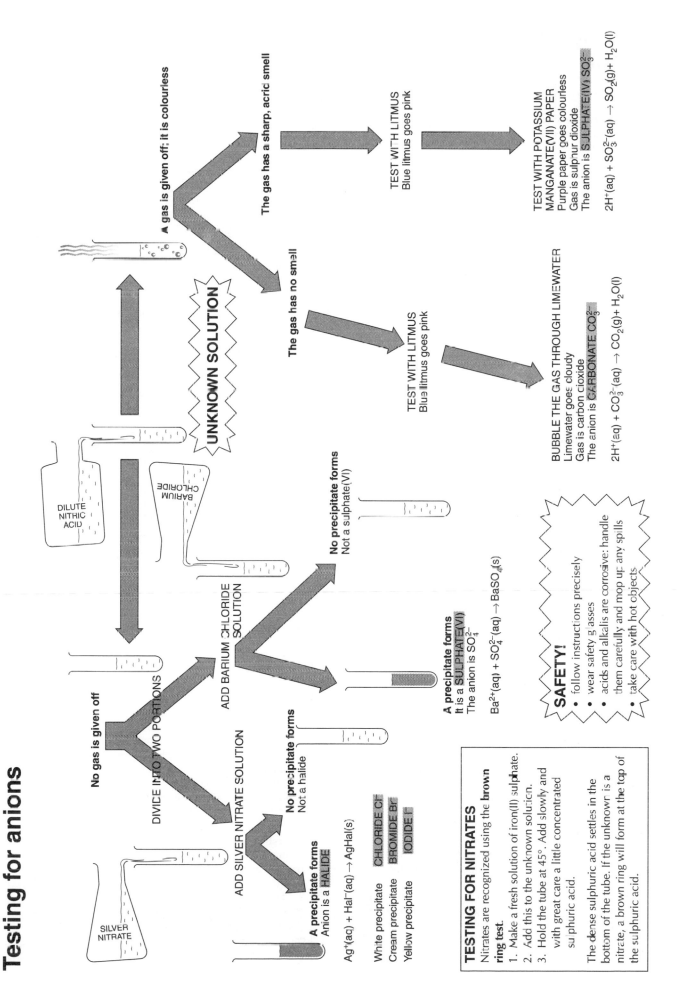

UNKNOWN SOLUTION

No gas is given off

DIVIDE INTO TWO PORTIONS

ADD SILVER NITRATE SOLUTION

SILVER NITRATE

A precipitate forms
Anion is a HALIDE

$Ag^+(aq) + Hal^-(aq) \rightarrow AgHal(s)$

White precipitate   CHLORIDE Cl⁻
Cream precipitate   BROMIDE Br⁻
Yellow precipitate   IODIDE I⁻

No precipitate forms
Not a halide

ADD BARIUM CHLORIDE SOLUTION

BARIUM CHLORIDE

No precipitate forms
Not a sulphate(VI)

A precipitate forms
It is a SULPHATE(VI)
The anion is SO_4^{2-}

$Ba^{2+}(aq) + SO_4^{2-}(aq) \rightarrow BaSO_4(s)$

DILUTE NITRIC ACID

A gas is given off; it is colourless

The gas has a sharp, acrid smell

TEST WITH LITMUS
Blue litmus goes pink

TEST WITH POTASSIUM MANGANATE(VII) PAPER
Purple paper goes colourless
Gas is sulphur dioxide
The anion is SULPHATE(IV) SO_3^{2-}

$2H^+(aq) + SO_3^{2-}(aq) \rightarrow SO_2(g) + H_2O(l)$

The gas has no smell

TEST WITH LITMUS
Blue litmus goes pink

BUBBLE THE GAS THROUGH LIMEWATER
Limewater goes cloudy
Gas is carbon dioxide
The anion is CARBONATE CO_3^{2-}

$2H^+(aq) + CO_3^{2-}(aq) \rightarrow CO_2(g) + H_2O(l)$

TESTING FOR NITRATES

Nitrates are recognized using the **brown ring test**.

1. Make a fresh solution of iron(II) sulphate.
2. Add this to the unknown solution.
3. Hold the tube at 45°. Add slowly and with great care a little concentrated sulphuric acid.

The dense sulphuric acid settles in the bottom of the tube. If the unknown is a nitrate, a brown ring will form at the top of the sulphuric acid.

SAFETY!

- follow instructions precisely
- wear safety glasses
- acids and alkalis are corrosive: handle them carefully and mop up any spills
- take care with hot objects

Making and testing for gases in the laboratory

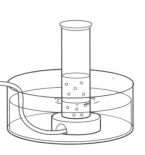

MAKING GASES

Gases are always made from the reaction between a solid and a liquid. The volume of gas made is controlled by the amount of liquid run into the flask holding the solid reactant.

Gas	Liquid	Solid	Reaction
Hydrogen	Dilute sulphuric acid	Magnesium	$H_2SO_4 + Mg \rightarrow H_2 + MgSO_4$
Oxygen	Hydrogen peroxide	Manganese(IV) oxide	$2H_2O_2 \rightarrow O_2 + 2H_2O$
Carbon dioxide	Dilute hydrochloric acid	Calcium carbonate	$2HCl + CaCO_3 \rightarrow CO_2 + CaCl_2 + H_2O$
Hydrogen chloride	Conc. sulphuric acid	Sodium chloride	$H_2SO_4 + 2NaCl \rightarrow 2HCl + Na_2SO_4$
Ammonia	Sodium hydroxide	Ammonium chloride	$NaOH + NH_4Cl \rightarrow NH_3 + NaCl + H_2O$
Sulphur dioxide	Dilute hydrochloric acid	Sodium sulphate(IV)	$2HCl + Na_2SO_3 \rightarrow SO_2 + 2NaCl + H_2O$
Chlorine	Conc. hydrochloric acid	Manganese(IV) oxide	$4HCl + MnO_2 \rightarrow Cl_2 + MnCl_2 + 2H_2O$
Nitrogen dioxide	Dilute nitric acid	Copper metal	$4HNO_3 + Cu \rightarrow Cu(NO_3)_2 + 2NO_2 + H_2O$

Drying gases

Because the liquid involved in the reaction is often an aqueous solution, the gas produced is often wet (contaminated with water).

Gases can be dried by passing them through something which reacts with (and so removes) the water.

Acidic and neutral gases are dried by passing them through concentrated sulphuric acid.

Ammonia is an alkaline gas. It is dried by passing it over calcium chloride granules.

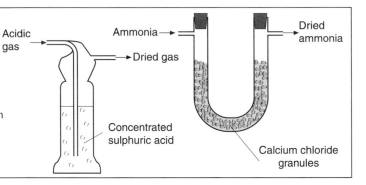

TESTING FOR GASES IN THE LAB

1. Look at its colour
Most gases are colourless but chlorine is green and nitrogen dioxide is brown.

4. Then confirm the identity of the gas with a special test
- hydrogen — ignites with a squeaky pop
- oxygen — relights a glowing splint
- carbon dioxide — turns calcium hydroxide (lime water) cloudy
- hydrogen chloride — white smoke forms with the ammonia bottle stopper
- ammonia — white smoke forms with the concentrated hydrochloric acid bottle stopper
- sulphur dioxide — potassium manganate(VII) on filter paper goes colourless
- chlorine — potassium iodide on filter paper goes brown
- nitrogen dioxide — a piece of copper goes green

2. Cautiously smell the gas
- Many gases have no smell: oxygen, hydrogen, nitrogen, carbon dioxide.
- Some smell acrid (this means a sharp smell which makes you jerk your head back): sulphur dioxide, hydrogen chloride.
- Others are pungent (they have a very strong smell): ammonia.
- Others have a characteristic or familiar smell: chlorine has a smell we all know from swimming pools.

3. Test with damp litmus paper
Remember that the only positive test is one in which the paper changes colour.
- Litmus goes pink (acidic): hydrogen chloride, sulphur dioxide, carbon dioxide.
- Litmus goes blue (alkaline): ammonia.

Indicator paper folded down the middle and wetted

INDEX

Entries in **bold** type indicate main topic entries.

cross-linking 54
crude oil 49, 50, 51
crust (of Earth) 7, 42, 43, 45
cryolite 46
crystals 5, 13, 31, 42, 63, 64
crystallization 31, 42

D

decanting 9
decomposition 22
delocalized electrons 13
denitrifying bacteria 41
density 18, 20, 43
deposition 42
destructive plate margin 43
detergents 27, 28, 36
deuterium 10
diamond lattice 16
diatomic molecules 12, 20
diffusion 5
digestion 36, 54
discontinuous rate method 34
displacement reactions 22
distillation 9
double bonds 15
ductility in metals 13
dynamic equilibrium 33

E

Earth, structure 43
Earth's crust, composition 7
Earth's magnetism 43
earthquakes 43
electrical properties 13, 14
electricity 5, 7, 16, 44, 47
electrode 62
electrolysis 22, 62
    of brine 48
    of molten sodium chloride 48
electrolysis calculations 62
electron 10–20, 62
    gain 32, 48
    loss 32, 48
    shells 10, 11
electronic configuration 10
electronic structure 10
electrostatic force 13–15
electrovalent bonds 14
element 7
empirical formula 56
endothermic changes 23
energy 5, 10, 13, 23, 35–40, 52
enthalpy 23
enzyme reactions 36
equations, chemical 22
equilibrium 33
erosion 42
esters 53
ethane 50
ethanoic acid 53
ethanol 52
    from fermentation 52
    industrial manufacture 52
    partial oxidation 52

ethene 51
    polymerization 51
    reaction with bromine 51
    reaction with hydrogen 51
    reaction with water 51
    uses 49
eutrophication 39
evaporation 9
evaporites 42
exothermic changes 23
extraction
    of chemicals from oil 49
    of metals 25, 46–48
extrusive igneous rocks 42

F

factors affecting rate of reaction 35
Faraday 62
fats and oils 51
faults 43
fermentation 36, 52
fertilizers 39
filtering 9, 31, 27, 37, 39, 63, 68
filtrate 9
fire triangle 23
fires, putting out 23
fixation of nitrogen 39
flame tests 66
fluoride 20
fluorine 10, 13, 17, 20
folds (in rocks) 42
food preservation 39, 48
forces between molecules 12
formulae of substances 7, 22, 50–52, 56
fossil fuels 38, 41, 49
fractional distillation 7, 8, 9
    of air 39
    of crude oil 49
fractions 8, 49
fractions of crude oil 49
    properties 49
    uses 49
freezing 6, 27, 39, 42
freezing point 6
fuels 23
fullerenes 16

G

gaining electrons 32
galvanizing 25, 45, 47
gases
    making in the lab 68
    tests for 68
giant covalent substances 16
glass 44
global warming 38
glucose 54
gneiss 42
gold 25
grain size 13
granite 42
graphite 16
greenhouse effect 38

group 1 17, 18, **19**
    compounds 19
    ions, tests for 66
group 2 17, 18, **19**
    compounds 19
    ions, tests for 66
    metal structure 17

H

Haber process 39, 40
haematite 42, 45
halide ions, tests for 20, 67
Hall cell 46
halogens 20
    reactivity 20
    structure 17
    uses 20
hardening of oils 51
hardness of metals 13, 17, 19, 47
hardness of water 27, 28
heat 5, 6, 23, 40
    of neutralization 30
heat treatment 13
helium 20
herbicides 20
heterogeneous mixtures 8
homogeneous mixtures 8
homologous series 50, 51
hydrocarbons 7, 8, 23, 49, 52, 56
hydrochloric acid 26
hydrogen
    making in the lab 68
    test for 68
    uses 48
hydrogen chloride
    making in the lab 68
    reactivity 25
    solution 26
    structure 17
    test for 68
hydrogen peroxide 34, 35, 68
hydroxide ion 19, 26, 30, 48

I

igneous rocks 42
immiscible liquids 9
incomplete combustion 23
indicator 29, 30, 53, 61, 63, 65
insoluble salts 31
insulator 7, 14–16
intrusive igneous rocks 42
iodide ion, test for 67
iodine lattice 16
ion 12
ion exchange 28
ionic bonding 14
ionic lattice 13, 16
ionic properties 14
ionic structure 14
iron extraction 45
iron, uses 25
iron(II) ion, test for 66
iron(III) ion, test for 66

reduction 32
relative atomic mass 55
relative formula mass 55
residue 9
respiration 38, 41
reversible reactions 33
rivers 27, 39
rock cycle 42
rock salt 42, 48
room temperature and pressure (r.t.p.) 58
r.t.p. 58
rust prevention 45
rusting 45

S
sacrificial protection 45
salt, chemicals from 48
   uses 48
salts 30
   making in the laboratory 63
   solubility 63
   in solution 26
sand 27, 43, 44
saturated compounds 49, 50
schist 42
sea-floor spreading 43
sea-water salts 7
sedimentary rocks 42
sediments 42
semiconductors 18
semi-metals 18
separating mixtures 9
sewage treatment 27
shale 42
shared pairs of electrons 15
shells of electrons 10
silica lattice 16
silicon 7, 17, 18
silicon dioxide 16
silver 20, 24, 31, 44, 47
slag 44, 45
smoke 44
soaps 27, 28
sodium
   reaction with water 19
   uses 25
sodium chloride 7, 8, 26, 48, 57
sodium hydroxide, uses 48
softening water 27, 28
soil 30, 39
solar energy 27
solar radiation 38
solubility 27, 28
   curve 28, 64
   of group 1 compounds 19
   measuring 64
solute 8
solutions 8
solvent 8
stability of group 1 and group 2
   compounds to heat 19
stainless steel 47
standard temperature and pressure (s.t.p.)
   58
starch 54
state symbols 22
states of matter 5, 6, 8, 9, 17

steam 6, 19, 25
steel 47
   alloy 47
   high carbon 47
   manufacture 45
   mild 47
   stainless 47
stomach acids 30
s.t.p. 58
strata 42
strong and weak acids 29
strong and weak alkalis 29
structure and boiling point 50
subduction 43
sublimation 5
sulphate(IV) ion, test for 67
sulphate(VI) ion, test for 67
sulphur chemistry 21
sulphur
   properties 21
   structure 17
sulphur dioxide 21
   making in the laboratory 68
   test for 68
sulphuric acid 21
supply/demand problem 49
surface area and rate of reaction 35
suspensions 8
system and surroundings 23

T
tectonic plates 43
temperature 5
   and enzymes 36
   and rate of reaction 35
temporary hardness 28
tests for gases 68
tests for ions 31, 66, 67
tests for water 26
thermal stability of group 1 and group 2
     compounds 19
thermosetting polymers 54
thermosoftening polymers 54
titanium 47
   extraction 47
   uses 47
titration 63, 65
transition elements 18, 47
   uses 47
transport and erosion 42
triatomic molecules 12
triple bonds 15
tritium 10

U
universal indicator 29, 53
unsaturated compounds 49

V
van der Waals' forces 12
vanadium(V) oxide 21
vapour 9, 21, 37
vinegar 53

viscosity 49
volatile compounds 49
volcanoes 43

W
water 26
   as a resource 27
   bonding and structure 26
   physical properties 26
   purification 26
   solvent properties 26
   tests for 26
   uses 27
   vapour 37
water cycle 27
water softening 27
water treatment 27
weak acids 29
weak bases 29
weathering 42
wind 5, 27, 42
xenon 20

Y
yeast 52
yoghurt 36

Z
Ziegler catalyst 51
zinc, uses 25
zinc(II) ion, test for 66

72